This book is to be returned on
or before the date stamped below

UNIVERSITY OF PLYMOUTH

PLYMOUTH LIBRARY

Tel: (0752) 232323
This book is subject to recall if required by another reader
Books may be renewed by phone
CHARGES WILL BE MADE FOR OVERDUE BOOKS

Application of Admixtures
in Concrete

OTHER BOOKS ON CONCRETE AND CONSTRUCTION MATERIALS AVAILABLE FROM E & FN SPON

Admixtures for Concrete: Improvement of Properties
Edited by E. Vazquez

Autoclaved Aerated Concrete: Properties, Testing and Design
RILEM Technical Committees 78-MCA and 51-ALC

Cement-based Composites: Materials, Mechanical Properties and Performance
A. M. Brandt

Concrete Admixtures
V. H. Dodson

Concrete Admixtures
P. Russell

Concrete in the Marine Environment
P. K. Mehta

Concrete Mix Design, Quality Control and Specification
K. W. Day

Construction Materials: Their Nature and Behaviour
Edited by J. M. Illston

Creep and Shrinkage of Concrete
Edited by Z. P. Bazant and I. Carol

Fly Ash in Concrete: Properties and Performance
Edited by K. Wesche

High Performance Concrete: From Material to Structure
Edited by Y. Malier

Hydration and Setting of Cements
Edited by A. Nonat and J. C. Mutin

Manual of Ready-Mixed Concrete
J. D. Dewar and R. Anderson

Special Concretes: Workability and Mixing
Edited by P. J. M. Bartos

Structural Grouts
Edited by P. L. J. Domone and S. A. Jefferis

For details of these and other books, contact The Promotions Department, E & FN Spon, 2-6 Boundary Row, London SE1 8HN, UK. Tel: 071-865 0066.

RILEM REPORT 10

Application of Admixtures in Concrete

State-of-the-Art Report prepared by RILEM Technical Committee TC 84-AAC, Application of Admixtures in Concrete

RILEM – The International Union of Testing and Research Laboratories for Materials and Structures

Edited by

A. M. Paillère
Laboratoire Central des Ponts et Chaussées, Paris, France

E & FN SPON
An Imprint of Chapman & Hall

London · Glasgow · Weinheim · New York · Tokyo · Melbourne · Madras

Published by E & FN Spon, an imprint of
Chapman & Hall, 2-6 Boundary Row, London SE1 8HN, UK

Chapman & Hall, 2-6 Boundary Row, London SE1 8HN, UK

Blackie Academic & Professional, Wester Cleddens Road, Bishopbriggs, Glasgow G64 2NZ, UK

Chapman & Hall GmbH, Pappelallee 3, 69469 Weinheim, Germany

Chapman & Hall USA, One Penn Plaza, 41st Floor, NY 10119, USA

Chapman & Hall Japan, ITP-Japan, Kyowa Building, 3F, 2-2-1 Hirakawacho, Chiyoda-ku, Tokyo 102, Japan

Chapman & Hall Australia, Thomas Nelson Australia, 102 Dodds Street, South Melbourne, Victoria 3205, Australia

Chapman & Hall India, R. Seshadri, 32 Second Main Road, CIT East, Madras 600 035, India

First edition 1995

© 1995 RILEM

Printed in Great Britain by St Edmundsbury Press, Bury St Edmunds, Suffolk

ISBN 0 419 19960 8

A catalogue record for this book is available from the British Library

Contents

Preface

Since it was formed in 1947, RILEM (The International Union of Testing and Research Laboratories for Materials and Structures) has been contributing to continuous transfer and exchange of knowledge and technologies, internationally. Its policy of active cooperation with the scientific and technical community, as well as with the industrial world, through promotion of applied and experimental research, has always been pursued by RILEM to respond to the emerging problems of modern society.

One of RILEM's principal activities is realised through its technical committees: the creation of Technical Committee 84-AAC on Applications of Admixtures for Concrete was approved during the RILEM General Council meeting in Tsukuba, Japan, in September 1984 and its chairman was appointed at that time. The rest of the committee members were chosen following the guidelines of RILEM.

The installation and first meeting of the Committee was held during March 1985 in Monterrey, Mexico. Later, the Committee met on nine occasions: Paris, France in July 1985; Madrid, Spain in April 1986; Treviso, Italy in March 1987; Paris, France in September 1987; Istanbul, Turkey in April 1988; Mexico City and Merida, Mexico in December 1988; Haifa, Israel in March 1989; Barcelona, Spain in May 1990, and Salzburg, Austria in April 1992.

The tasks of Committee 84-AAC were:

1. Definition of admixtures:
Admixtures for concrete, mortar or paste are inorganic (including minerals) or organic materials in solid or liquid state, added to the normal components of the mix, in most cases up to a maximum of 5% by mass of the cement or cementitious materials.

The admixtures interact with the hydrating cementitious system by physical, chemical or physico-chemical action, modifying one or more properties of concrete, mortar or paste, in the fresh, setting, hardening or hardened state.

Materials such as fly ash, slag, pozzolanas, or silica fume which can be constituents of cement and/or concrete, also products acting as reinforcement, are not classified as admixtures.

2. Preparing a Guide for the use of admixtures
The Guide was published in the RILEM Journal, *Materials and Structures*, Vol. 25, January 1992, pp. 49-56, and is also included as an Appendix to this Report. It includes terminology, definitions of the properties of concrete that can be affected by the use of admixtures and other terms related to admixtures.

3. Preparation of a state-of-the-art of concrete admixtures
This is presented in this volume.

4. Compilation of a bibliography

5. Organization of an international symposium
The symposium 'Admixtures for Concrete: Improvement of Properties' was held in Barcelona, Spain in May 1990 and the proceedings volume was published by Chapman & Hall.

Following the introductory chapter on the interaction of admixtures in the cement–water system, this volume is divided into two parts, the first part covering the most commonly used admixtures and the second part those not so commonly used.

The tasks of the Committee were accomplished by all its members even though an individual member was designated to prepare a specific paper; all the subjects were discussed and approved paragraph by paragraph by the Committee.

I must express on behalf of RILEM and myself my appreciation and special thanks to all members of this Committee for the effort and hard work to accomplish the tasks of RILEM TC 84-AAC.

The most sincere gratitude is also offered to all the organizations which allowed a member of their staff to be part of this Committee and provided them with all the facilities and support with which all the works assigned to the Committee were accomplished.

Prof. Dr. Ing. R. Rivera-Villarreal
Chairman of TC 84-AAC

RILEM Technical Committee TC 89-AAC

This state of the art was drafted within the framework of RILEM Technical Committee 89-AAC *Application of Admixtures for Concrete*, which was comprised as follows:

Chairman
Dr R. Rivera Villareal, Universidad Autonoma de Nuevo Leon, Mexico

Secretary
Dr A. M. Paillère, Laboratoire Central des Ponts et Chaussées, Paris, France

Members
Dr S. Akman, Technical University, Istanbul, Turkey
Dr F. Alou, EPFL, Lausanne, Switzerland
Dr M. Ben Bassat, Technion, Haifa, Israel
Dr S. Biagini, MAC, Treviso, Italy
E. Decker, Consulting Engineer, Roanoke, USA
Dr D. Dimic, Institute for Research and Testing of Materials and Structures, Ljubljana, Slovenia
Dr F. Massazza, Italcementi, Bergamo, Italy
Dr S. Nagataki, Tokyo Institute of Technology, Japan
Dr V. S. Ramachandran, National Research Council, Ottawa, Canada
Dr E. Vazquez, Universidad Politecnica de Catalunya, Barcelona, Spain

1

Interaction of admixtures in the cement-water system

V. S. Ramachandran

1.1 ABSTRACT

Physical adsorption, chemisorption and chemical interactions occur between admixtures (such as accelerators, retarders, water reducers and superplasticizers) and hydrating cement. The mechanism of action of admixtures, changes in water demand, microstructure, strength and durability of fresh and hardened cement can be explained by the interaction effects.

1.2 INTRODUCTION

Admixtures confer certain beneficial effects to concrete, including reduced water requirement, increased workability, controlled setting and hardening, improved strength and better durability.

Many approaches have been adopted to investigate the role of admixtures. One approach is to determine the state of the admixture in concrete at different times of curing. The admixture may remain in a free state as a solid or in solution, or interact at the surface or chemically combine with the constituents of cement or cement paste. The type and extent of the interaction may influence the physico-chemical and mechanical properties of cement. In this chapter an attempt is made to discuss the possible interactions of different types of admixtures in the cement-water system, with particular reference to the changes in physico-mechanical properties.

1.3 DISCUSSION

1.3.1 Accelerators

Many inorganic compounds such as chlorides, fluorides, carbonates, silicates, aluminates, borates, nitrites, thiosulphates, triethanolamine, diethanolamine, and formates have been advocated for use as accelerators.

Calcium chloride is perhaps the most efficient and economical accelerator. In the hydration of C_3S ($C = CaO$, $S = SiO_2$, $H = H_2O$) there is evidence that calcium chloride exists in different states in the C_3S paste. Based on thermal analysis and leaching studies, Ramachandran [1] has concluded that, depending on the time of hydration, the chloride may exist in a free form (extractable by ethyl alcohol), incorporated strongly into the C-S-H phase (unleachable with water), chemisorbed or in interlayer positions (leachable with water). At 168 h about 20% of the added chloride is incorporated strongly into the C-S-H phase. The rest is in a chemisorbed state in the interlayers (easily leachable with water). These results may have implications in explaining the mechanism of the accelerating action, microstructural development, differences in the C/S ratio of C-S-H products, corrosion potential of chlorides and intrinsic properties of Portland cement. Calcium chloride accelerates the reaction between C_3A and gypsum, and monochloroaluminate is formed after all gypsum is consumed [2]. Using leaching and pressure techniques it has been found that more chloride is immobilized by the C_3A + gypsum mixture than by C_3S or Portland cement [3].

(a) Triethanolamine
Triethanolamine is termed an accelerator. It acts, however, as a retarder of C_3S hydration [4]. An examination of the thermal behaviour of C_3S hydrated to different periods in the presence of triethanolamine reveals the development of exothermal peaks that could be attributed to the decomposition of a complex of the amine with the hydrated products of C_3S. This complex may be responsible for early retardation and possibly for the higher C/S ratio of the C-S-H product. Triethanolamine accelerates the reaction between C_3A and gypsum [5]. The cation-active amine may react with aluminium and calcium ions on the surface of C_3A. In cements, at a

dosage of 0.1 or 0.5% triethanolamine, setting occurs rapidly within about 2–6 min; this is attributed to the accelerated formation of ettringite and C_3A-hydration products [5].

(b) Formates
Calcium formate is used as a non-chloride accelerator. In cements more ettringite is formed with formate than with calcium chloride [6]. The increased formation of ettringite in the presence of calcium formate is attributed to the formation of the complex $C_3A.3Ca(HCO_2)_2.30H_2O$ at ordinary temperatures, analogous to ettringite. Enhanced formation of ettringite in the presence of calcium acetate and calcium propionate is also explained by their ability to form hexagonal prism-type phases with C_3A [7]. Calcium formate accelerates the hydration of the C_3S surface [8].

(c) Other accelerators
Sodium carbonate decreases the setting time of cement by 2–4 h but after 10–12 h the hydration is retarded due to the precipitation of $CaCO_3$ by the reaction of Na_2CO_3 with lime. The precipitation occurring within the pores of the product decreases permeability [9]. Oxalic acid may also act as an accelerator by decreasing the setting time of cement by 43% and increasing strengths by 12% [10]. Strength development is explained by the formation of calcium oxalate, although the amount formed may be small.

Although calcium chloride is a comparatively simple molecule compared to other accelerators, its accelerating mechanism is not resolved. At least 12 theories have been proposed for its action [11]. It can therefore be appreciated that the role of more complex compounds involving organic compounds is not easily resolved.

1.3.2 Retarders

Organic compounds such as unrefined Na, Ca or NH_4 salts of lignosulphonic acids, hydroxy-carboxylic acids, carbohydrates and inorganic compounds (oxides of Pb and Zn phosphates, Mg salts, fluorates and borates) act as retarders.

(a) Sugars

Theories of retarding effect are based on adsorption, precipitation, complexation or nucleation. In all these processes, interactions are involved. According to Milestone, sugar and sugar acids adsorb on to Ca^{2+} ions on the hydrating C_3S surface and poison the C–S–H nucleating sites [12]. Adsorption of glucose, for example, results in the zeta potential becoming negative from the positive values. Poisoning of CH nuclei by adsorption of sugars is also envisaged [13].

According to the precipitation theory, addition of mono or polysaccharides increases the concentration of Ca, Al and Fe. Sugars may combine with them to form insoluble metal organic complexes on the cement grains and retard hydration [14]. All sugars do not retard cement to the same extent.   Non-reducing sugars, for example containing five-membered rings (sucrose, raffinose), are the best retarders. Although no sucrose-silicate or calcium complex has been detected with these sugars, it is suggested that a half salt formed by attachment of Ca and OH groups to the five membered ring may poison CH and C–S–H nuclei [15]. Retardation of C_3A to the cubic C_3AH_6 is explained by the formation of interlayer complex of the hexagonal aluminate hydrate with the organic compounds [16, 17].

It appears that no single theory can be applied to explain the behaviour of all sugars on all cement components under all conditions of hydration.   Adsorption need not occur only on unhydrated or hydrated surfaces.   Some sugars accelerate the initial formation of ettringite in the C_3A–gypsum–H_2O system and others retard it. Sucrose is an accelerator, whereas raffinose and trehalose are retarders. When sucrose is used, it is assumed that adsorption occurs on the anhydrous surface, preventing the formation of an impermeable layer of ettringite [18]. Why this does not occur with other sugars needs to be examined.

The precipitation theory cannot be applied for all cases.   For example, the stability constants of calcium complexes of various compounds do not bear correlation with their potency as retarders. A study of many complexes has shown that some are retarders and others are not. The theory based on poisoning of $Ca(OH)_2$ nuclei by itself may not always cause retardation. Although $Ca(OH)_2$ growth can be modified by incorporation of dyes it does not show any retarding characteristics [18].   Nucleation effect of C–S–H should also be considered. A combination of two mechanisms may occur and it is not easy to separate the occurrence of two mechanisms. For example, EDTA retards the hydration of C_3S. The retarding action may be explained by the precipitated gel coating of C–S–H on the hydrating C_3S, consequent on the formation of a complex between EDTA and $Ca(OH)_2$ in the solution phase [19]. The idea that retarding agents should contain the α-hydroxy carbonyl group could be questioned. According to Daugherty and Kowalewski [20], organic compounds with two or more (OH) groups are necessary for the retardation of C_3A hydration.

(b) Hydroxycarboxylic acids
Adsorption studies have been carried out on hydroxycarboxylic acids such as salicylic acid on cement and cement components [21–23]. Only a small amount of adsorption occurs on the unhydrated phases compared to that on hydrated products of cement minerals. A complex of salicylic acids with aluminium may be responsible for the retardation of hydration of C_3A contents. At lower C_3A contents, smaller amounts of retarder are adsorbed leaving larger amounts of the admixture to affect the C_3S component. Alkalis may affect dissolution and interaction reactions.

Early hydration of C_3A + gypsum + CH may be accelerated and later reaction involving the conversion of ettringite to monosulphoaluminate may be retarded by citric acid [24]. Acceleration in the initial states may be due to preferential adsorption of citric acid which promotes hydrolysis of C_3A to hexagonal phases. The later retardation may be due to the formation of a complex between citric acid and monosulphate. Calcium citrate formed by reaction of lime and $CaCO_3$ (impurities) with citric acid may hinder the development of nuclei in the hydration of plaster of paris.

Correlation between solubility and retarding behaviour is not always possible. Solubility of oxalic acid is 5×10^{-6} M and that of gluconic acid is 8×10^{-5} M. Oxalic acid is not a retarder for C_3A hydration although less soluble, but gluconic acid is more soluble and is a good retarder [25].

(c) Lignosulphonates
The relative retarding effects of sugars and lignosulphonate are not easy to resolve. The hydrating C_3A adsorbs irreversibly substantial amounts of lignosulphonate and hence it is possible that pure

lignosulphonate contributes to the retarding effect.

Thermograms show that both commercial and sugar-free lignosulphonates are equally effective in retarding the hydration of cement [26]. Both the commercial and pure lignosulphonates retard setting times. According to other investigations either the sugar-free lignosulphonate is a poor retarder or inert in its action on cement. The disagreement may be due to the difficulty of preparing pure lignosulphonate and the differences in its molecular weights.

(d) Inorganic retarders

Many inorganic salts retard the hydration of cement. These salts form insoluble hydroxides in alkaline solution and may form a coating on the cement particles. There is evidence that inorganic compounds form complexes with the hydrating cement hydration. Zinc oxide retards the hydration of the C_3S and does not influence the hydration of C_3A + gypsum mixture. The formation of calcium hydroxyzincate Ca $(Zn(OH)_3H_2O)_2$ by the reaction of the $Ca(OH)_2$ with ZnO has been confirmed [27]. At an addition of 10% Zn equivalent of hydroxyzincate, 3.3% Zn may be incorporated into the C–S–H phase. The retardation effect of $Pb(NO_3)_2$ is attributed to the very rapid precipitation of a protective $Pb(OH)_2$ on the cement grains [28]. Most phosphates retard setting. The adsorption of phosphate ions at the surface of the clinker phase or on the first hydration product is thought to result in the precipitation of calcium phosphates.

1.3.3 Water reducers

Water reducers consist of Ca, Na or NH_4 salts of lignosulphonic acid, Na, NH_4 or triethanolamine salts of hydroxycarboxylic acid and carbohydrates. Lignosulphonates containing (OH), (COOH) and SO_3H groups are more widely used than others. Hydroxycarboxylic acids, such as citric acid, tartaric acid, salicylic acid, heptonic acid, saccharic acid and gluconic acid, contain (OH) and (COOH) groups. Gluconic acid-based admixtures are used extensively. Carbohydrates include glucose, sucrose or hydroxylated polymers obtained by partial hydrolysis of saccharides. The role of water reducers (normal, accelerating or retarding) in terms of their effect on hydration of cement is similar to that of retarders, accelerators and superplasticizers.

The effect of some accelerating and retarding admixtures has already been discussed. In this section the interaction of lignosulphonates with cement will be emphasized.

The plasticizing action of water reducers is related to their adsorption and dispersing effects in the cement–water system. In other words, some sort of interaction seems to be involved.

(a) Tricalcium aluminate

The hydration of tricalcium aluminate is retarded by lignosulphonate. The interaction between lignosulphonate and hydrating C_3A can be studied by adsorption experiments. Adsorption isotherms cannot be obtained in the C_3A–lignosulphonate–H_2O system because hydration of C_3A occurs during the measurements, especially at low admixture concentrations. It is possible to determine the adsorption–desorption isotherm in the system hexagonal aluminate–lignosulphonate–H_2O. Scanning loops in the isotherms show complete irreversibility, indicating a complex formation. Increase in the c-axis spacing of the hexagonal phase containing lignosulphonate is caused by the formation of an interlayer complex. The interlayer complex would impede the conversion of the hexagonal phase to the cubic phase. In a non-aqueous phase, C_3A does not adsorb any lignosulphonate. The retarding effect seems to be due to the reaction of hydrating C_3A and lignosulphonate. Jawed *et al.* [29] observed more fluidity in the cement paste containing a mixture of lignosulphonate and Na_2CO_3 than when each of them was used separately. They proposed that an ionic complex occurred between lignosulphonate and CO_3 and it was more anionic than lignosulphonate and, hence, acted as a better dispersant.

The conversion of ettringite to monosulphoaluminate is retarded by adding a water reducer to C_3A + gypsum. One of the suggested mechanisms of the plasticizing action may be related to a lower water demand caused by the retardation of ettringite formation and decrease in the interlocking of the ettringite particles. This observation has to be confirmed by a more extensive investigation.

(b) Tricalcium silicate

The hydration of tricalcium silicate is retarded by lignosulphonate. Apparent adsorption–desorption isotherms of the system C_3S–lignosulphonate–H_2O can be explained by dispersion and hydration effects [30]. There is only partial reversibility during desorption, indicating the existence of a strongly bound surface complex involving C_3S, lignosulphonate and H_2O. Such a complex may cause retardation of hydration of C_3S. That the hydrated C_3S paste adsorbs irreversibly substantial amounts of lignosulphonate may be concluded from adsorption isotherms on the completely hydrated C_3S. Even in the non-aqueous medium the hydrated C_3S adsorbs calcium lignosulphonate, unlike the anhydrous C_3S phase. These results suggest that the retarding effect involves a reaction of lignosulphonate with the hydrating C_3S surface.

(c) Tricalcium aluminate – tricalcium silicate

The C_3A phase adsorbs larger amounts of lignosulphonate than C_3S when exposed to aqueous solutions of lignosulphonate. Therefore, in cements hydrating C_3A may act as a sink for lignosulphonate. It is known that when lignosulphonate is added a few minutes after water has come into contact with cement, the hydration of C_3S is retarded more strongly. This would mean that the hydrated C_3A phase adsorbs less lignosulphonate and leaves larger amounts of the admixture in the solution phase for retardation of C_3S hydration.

The effect of lignosulphonate on cement depends not only on the amounts of C_3A and C_3S but also on the alkalis, SO_3, particle size of cement, etc. Depending on these factors and others early set may be retarded or accelerated but the final set is generally retarded.

The early acceleration of set is prompted in cements with higher aluminate/SO_3 ratios. At early times, due to the adsorption of lignosulphonate on C_3S, $Ca(OH)_2$ is not released and the rate of formation of ettringite is increased. This implies that C_3A + gypsum reaction to form ettringite is faster than that containing $Ca(OH)_2$.

It is apparent that the role of the cationic lignosulphonate molecule as a water reducer is related to its adsorption on the cement components. The amount of adsorption, the rate and how strongly it is adsorbed may play a role in the setting, dispersibility, microstructure, shrinkage, durability and other properties of cement. The effect of cations and the material weight of lignosulphonate on cement properties has not been explored fully.

1.3.4 Superplasticizers

Most superplasticizers are based on sulphonated melamine formaldehyde (SMF), sulphonated naphthalene formaldehyde (SNF) and modified lignosulphonates. The action of water reducers involves adsorption and dispersion in the cement–water system. Similarly, effective surface interaction and dispersion of cement occur when superplasticizers are used. A study of the rate and amount of adsorption of superplasticizers on cement and cement compounds has provided some information on the rheological, setting and hydration characteristics. The general results of interaction and adsorption characteristics of SMF are similar to those of SNF.

(a) Tricalcium aluminate
The hydration of C_3A in the presence of SMF indicates that hydration is retarded. Adsorption of SMF on C_3A occurs as soon as the solution comes into contact with it [31]. The amount and rate of adsorption on the C_3A phase far exceed those on C_3S or Portland cement. In a non-aqueous medium adsorption is nil on C_3A but small amounts of SMF are adsorbed by the hexagonal phase. Adsorption is irreversible, indicating that a chemical interaction occurs between the hydrating C_3A and SMF. Thus the retardation of C_3A may be explained by strong adsorption of SMF on the hydrating C_3A surface.

The reported results on the rate of hydration of C_3A + gypsum mixture containing superplasticizers are contradictory. Adsorption of SMF has been studied on the C_3A + gypsum system prehydrated for various periods. Desorption experiments show that SMF is irreversibly adsorbed. A surface chemical or chemical interaction seems to occur between the hydrating C_3A or C_3A-gypsum mixture with SMF. The enhanced dispersion effect of superplasticizer when it is added a few minutes after mixing water is added to concrete can be explained as outlined in section 1.3.3.

(b) Tricalcium silicate
Hydration of C_3S is retarded by SMF. Irreversible adsorption is indicated in the system containing C_3S-SMF-H_2O. This may involve interaction between hydrating C_3S and superplasticizer. The interaction of superplasticizers with hydrating C_3S, C_3A and C_3A + gypsum components cannot be directly applied to the cement

behaviour. The interaction within the cement system is much more complex because, in addition to interfering effects of the silicates and aluminates, the alkalis and SO_3 also play an important role.

(c) Cement

The amount of adsorption of SMF on cement varies with length of exposure to the solution. Within a few seconds there is a steep increase in adsorption due to the C_3A–C_4AF components in the cement. Further adsorption does not occur up to about four to five h, after which it increases continuously. Adsorption beyond about five h is mainly due to the hydrating C_3S component. These results suggest that adsorption and interaction of SMF with the cement components are involved in the dispersion of cement and retardation of cement hydration.

The amount of adsorption of superplasticizer on cement can be related to workability. With the SNF superplasticizer the mini slump values increase as the amount of adsorption increases [32]. The adsorption values of SNF on three types of cement decrease as follows:

Type III > Type I > Type II [33].

The C_3A/SO_3 ratios in the cements follow the same trend. That adsorption is dependent on C_3A content becomes clear: for the same workability, a higher dosage of superplasticizer is required for Type I than for Type V cement. According to the ASTM designation, Types I, II, III, IV and V are described as, general purpose, moderate sulphate resistant-moderate heat of hydration, high early strength, low heat and sulphate resisting cements.

Zeta potential development in suspensions of cement, alite, C_3A and $Ca(OH)_2$ containing superplasticizers has been studied. It is generally found that both adsorption and zeta potential values increased as the concentration of superplasticizer added to cement is increased [34].

The mechanism of retardation of hydration and even dispersion may not be entirely due to the adsorption effect of the anions. In SNF superplasticizer containing NH_4, Co, Mn, Li and Ni cations, the time for the development of maximum heat is 12.7 h with NH_4, and only 9.25 h with Ni [35]. The relative roles of cations in the super-plasticizers are not well understood.

Higher-than-normal workability of concrete containing a superplasticizer is maintained for about 30–60 min after which the slump value decreases. In the period during which slump loss is occurring the C_3A phase reacts with gypsum. It is possible that the extent of reaction of C_3A and gypsum and the crystalline form of the product would have an important effect on the workability of concrete. Addition of superplasticizer enhances the initial reaction between C_3A and gypsum. Alkalis enhance this reaction. According to Hattori [36] coagulation of the particles plays a more important role than the chemical bonding in slump losses. Experiments on C_3S–SMF–H_2O system have shown that rapid loss of slump occurs in this system [37]. Thus the contribution of the C_3S phase should also be taken into account. All these mechanisms involve some sort of interaction with the superplasticizer.

One of the methods of maintaining the slump in superplasticized concrete is to add retarders such as calcium gluconate to cement [38]. It is possible that the retarder, although not interfering with dispersion caused by the superplasticizer, adsorbs on the cement components, affecting the chemical or physical processes that cause agglomeration or interlocking of the cement particles.

1.4 CONCLUSIONS

Admixtures in aqueous solutions or in the solid form may interact with the hydrating cement compounds in the cement–water–admixture system. Physical, chemical and mechanical properties of the cement paste are affected to different extents by these interactions. In many instances, conclusion on interaction processes are speculative because they are based on indirect evidence. A study of interaction occurring in systems containing more than one admixture becomes even more complex. Even the so-called single component commercial admixture may contain a small amount of a chemical that could interfere with the interaction of the main component. Disagreement in reported results on interactions in cement systems may be traced to the variability in the characteristics of the starting materials, methods of curing, testing methods and interpretation.

1.5 REFERENCES

1. Ramachandran, V. S., (1971) Possible states of chloride in the hydration of tricalcium silicate in the presence of calcium chloride, *Materials and Structures*, Vol. 4, pp. 3–12.
2. Tenoutasse, N., (1968) The hydration mechanism of C_3A in the presence of calcium chloride and calcium sulphate, *Proceedings of Fifth International Symposium on Chemistry of Cement*, Tokyo, pp. 372-8.
3. Ramachandran, V. S., Seeley, R. C., and Polomark, G. M., (1984) Free and combined chloride in hydrating cement and cement components, *Materials and Structures*, Vol. 17, pp. 285-9.
4. Ramachandran, V. S., (1972) Influence of triethanolamine on the hydration characteristics of tricalcium silicate, *Journal of Applied Chemistry and Biotechnology*, Vol. 22, pp. 1125–38.
5. Ramachandran, V. S., (1976) Hydration of cement – role of triethanolamine, *Cement and Concrete Research*, Vol. 6, pp. 623–32.
6. Bensted, J., (1978) Early hydration behaviour of Portland cement in water, calcium chloride and calcium formate solutions, Part I, *Silicates Industriels*, Vol. 43, pp. 117-22.
7. Bensted, J., (1980) Early hydration behaviour of Portland cement in water, calcium chloride and calcium formate solutions, Part III, *Silicates Industriels*, Vol. 45, pp. 5–10.
8. Singh, N. B. and Abha Km., (1983) Effect of calcium formate on the hydration of tricalcium silicate, *Cement and Concrete Research*, Vol. 13, pp. 619–25.
9. Collepardi, M., Marcialis, A. and Massidda, L., (1973) The influence of sodium carbonate on the hydration of cements, *Annali di Chimica*, Vol. 63, pp. 83–93.
10. Djabarov, N. B., (1970) Oxalic acid as an additive to cement, *Zement –Kalk –Gips*, Vol. 23, pp. 88–90.
11. Ramachandran, V. S., (1976) *Calcium Chloride in Concrete*, Applied Science Publishers, UK, p. 216.
12. Milestone, N. B., (1979) Hydration of tricalcium silicate in the presence of lignosulphonate, glucose and sodium gluconate, *Journal of the American Ceramic Society*, Vol. 62, pp. 321-4.
13. Young, J. F., (1972) A review of the mechanisms of set retardation in Portland cement pastes containing organic admixtures, *Cement and Concrete Research*, Vol. 2, pp. 415-33.

14. Suzuki, S. and Nishi, S., (1959) The effects of saccharides and other organic compounds on the hydration of cement, *Semento Gijutsu Nempo*, Vol. 13, pp. 160-70.
15. Thomas, N. L. and Birchall, J. D., (1983) The retarding action of sugars on cement hydration, *Cement and Concrete Research*, Vol. 13, pp. 830–42.
16. Young, J. F., (1968) The influence of sugars on the hydration of tricalcium aluminate, *Proceedings of Fifth International Symposium on the Chemistry of Cement*, Tokyo, pp. 256–67.
17. Young, J. F., (1970) Effect of organic compounds on the inter-conversions of calcium aluminate hydrates: Hydration of tricalcium aluminate, *Journal of American Ceramic Society*, Vol. 53, pp. 65–9.
18. Ramachandran, V. S., Feldman, R. F. and Beaudoin, J. J., (1981) *Concrete Science*, Heyden & Son, UK, p. 427.
19. Thomas, N. L. and Double, D. D., (1983) The hydration of Portland cement, C_3S and C_2S in the presence of a calcium complexing admixture (EDTA), *Cement and Concrete Research*, Vol. 13, pp. 391–400.
20. Daugherty, K. E., and Kowaleski, M. J., (1968) Effect of organic compounds on the hydration reactions of tricalcium aluminate, *Proceedings of Fifth International Symposium Chemistry of Cement*, Tokyo, Vol. 4, pp. 42–52.
21. Diamond, S., (1971) Interactions between cement minerals and hydroxycarboxylic retarders: I, Apparent adsorption of salicylic acid on cement hydrated cement compounds, *Journal of the American Ceramic Society*, Vol. 54, pp. 273–6.
22. Diamond, S., (1972) Interactions between cement minerals and hydroxycarboxylic acid retarders: III, Infrared spectral identification of aluminosalicylate complex, *Journal of the American Ceramic Society*, Vol. 55, pp. 405–8.
23. Rossington, D. R. and Runk, E. J., (1968) Adsorption of admixtures on Portland cement hydration products, *Journal of the American Ceramic Society*, Vol. 51, pp. 46-50.
24. Tinnea, J. and Young, J. F., (1977) Influence of citric acid on the reactions in the system $3CaO.Al_2O_3 - CaSO_4.2H_2O - CaO - H_2O$, *Journal of the American Ceramic Society*, Vol. 60, pp 387–9.
25. Ramachandran, V. S., (1972) Effect of lignosulphonate on tricalcium aluminate and its hydration products, *Materials and Structures*, Vol. 5, pp. 67–76.

26. Ramachandran, V. S., (1978) Effects of sugar-free lignosulphonates on cement hydration, *Zement –Kalk –Gips*, Vol. 31, pp. 206–10.
27. Lieber, W., (1967) Influence of zinc oxide on the setting and hardening of Portland cements, *Zement –Kalk –Gips*, V. 20, pp. 91–5.
28. Thomas, N. L., Jameson, D. A. and Double, D. D., (1981) The effect of lead nitrate on the early hydration of Portland cement, *Cement and Concrete Research*, Vol. 11, pp. 143-53.
29. Jawed, I., Klemm, W. A. and Skalny, J., (1979) Hydration of cement-lignosulphonate-alkali carbonate system, *Journal of the American Ceramic Society*, Vol. 62, pp. 461-4.
30. Ramachandran, V. S., (1972) Interaction of calcium lignosulphonate with tricalcium silicate, hydrated tricalcium silicate and calcium hydroxide, *Cement and Concrete Research*, Vol. 2, pp. 179–94.
31. Ramachandran, V. S., (1981) Influence of superplasticizers on the hydration of cement, *Proceedings of Third International Congress on Polymers in Concrete*, Koriyama, Japan, Vol. II, pp. 1071-81.
32. Collepardi, M., Corradi, M., and Valente, M., (1981) Influence of polymerization of sulphonated naphthalene condensate and interaction with cement, *American Concrete Institute Special Publication*, SP-68, pp. 485–98.
33. Ramachandran, V. S., (1984) *Concrete Admixtures Handbook: Properties, Science and Technology*, Noyes Publications, New York, p. 626.
34. Collepardi, M., Corradi, M., Baldini, A. and Pauri, M., (1980) Influence of sulphonated naphthalene on the fluidity of cement paste, *Proceedings of Seventh International Symposium on Chemistry of Cement*, Paris, Vol. III, pp. 20-5.
35. Sinka, J., Fleming, J. and Villa, J., (1978) Condensates on the properties of concrete, *Superplasticizers in Concrete*, American Concrete Institute, SP-62, pp. 22–36.
36. Hattori, K., (1978) Experiences with Mighty superplasticizer in Japan, *Superplasticizers in Concrete*, ACI SP-62, pp. 37–66.
37. Duston, C. J. and Young, J. F., (1982) *Influence of Superplasticizers on the Early Hydration of Portland Cement and Cement Compounds*, Report FHWA/U1-196, p 108, University of Illinois.
38. Ramachandran, V. S., (1981) Effect of retarders/water reducers on slump loss in superplasticized concrete, *American Concrete Institute SP-68*, pp. 393–407.

PART ONE

ADMIXTURES FOR CONCRETE
IN MOST FREQUENT USE

2

Air entraining admixtures

R. Rivera

2.1 DEFINITION

Air entraining admixtures cause the formation of fine uniformly distributed micro-bubbles for air in the concrete or mortar, and remain after hardening. Air entraining makes hardened mortars and concretes freeze-thaw resistant.

2.2 INTRODUCTION

Some air entraining agents used in cement paste or mortar are not considered as admixtures according to the RILEM admixtures definition, but as air entraining admixtures when they are incorporated in amounts higher than 5% of cementitious constituents.

These admixtures act by physical action. The air cells are usually added to the mixer as a stable preformed foam and thoroughly blended into the mix. The air cell may also be made mechanically by entrapping air during high speed mixing of the cement paste or mortar materials, when a surface active agent is added.

The air bubbles produced by air entraining admixtures are dispersed throughout the cement paste due to surface active properties of these admixtures. The bubbles are separate from the capillary pore system in the cement paste and they never become filled with the products of hydration of cement as gel can form only in water [p. 291 of 1].

The improved resistance of air entrained concrete to frost attack was discovered accidentally in the late 1930s when it was observed that concrete pavements in New York State made with cement that has been manufactured with grinding aids, that included beef fat, calcium stearate and fish oil, were more durable than others and had acted apparently as air entraining agents [2].

Air entraining improves the workability and consistency of fresh concrete and reduces its segregation and bleeding. Therefore, it is possible to take advantage in proportioning, using less water and sand. The strength of the hardened concrete is decreased as the amount of air is increased, but the effects are compensated by the appropriate proportioning changes.

2.3 COMPOSITION

Air entraining agents belong to a class of chemicals called surfactants, which is a short for surface active substances, whose molecules are absorbed strongly at air-water or solid-water interfaces. This behaviour is due to a specific molecular property that is called amphipathic [3].

The essential requirements of an air-entraining agent are that it rapidly produces a system consisting of a large number of stable finely divided small voids uniformly distributed through the cement paste, that the individual bubbles must resist coalescence, and that the foam must have no harmful chemical effect on the cement.

Numerous proprietary brands of air entraining agents are available commercially [3, 4, 5]. They have been categorized in different ways [1]. One of them is [6] as follows:

1. Salts and wood resins (vinsol);
2. Synthetic detergents;
3. Salts of petroleum acids;
4. Salts of proteinaceous materials;
5. Fatty and resinous acids and their salts;
6. Organic salts of sulphonated hydrocarbons.

A list of some suppliers is given in Table 5.2 on p. 273 of [7] showing brand names and some of the properties of the admixtures. The performance of the unknown ones should be checked by trial mixes.

Most of the commercial air-entraining admixtures are available in liquid state, although a few are powder, flakes or semi-solids.

2.4 MECHANISM OF ACTION

For each concrete mix there is a minimum volume of voids required for protection from frost. It was found [8] that this volume corresponds to 9% of the volume of mortar and it is essential that the air bubbles are distributed throughout the cement paste. The adequacy of air entrainment can be estimated by the spacing factor. This factor is the maximum distance of any point in the cement paste from the periphery of a nearby air void [9]. A maximum spacing factor of 0.20 mm is required for full protection from frost damage [10]. The smaller the spacing factor, the better the durability of concrete [11]. The voids are not all of the same size, and range usually up to 300 μm. If bubbles are bigger that 300 μm the air entrained is not effective [p. 292 of 7].

2.5 MAIN FACTORS INFLUENCING THE MECHANISM OF ACTION

2.5.1 Amount of air

Generally the larger the quantity of the admixture, the more air is entrained, but there is a dosage limit beyond which there is no further increase in the volume of voids [pp. 293, 295 of 1].

There are some other aspects influencing the amount of air actually entrained in concrete when a given amount of air entraining admixture is added, due to variations in its ingredients, such as cement type, supplementary cementing materials, other admixtures, and mixing, handling and temperature:

1. Smaller amounts of air entraining admixtures are generally required as the fineness of the cementitious products increase.
2. Alkali content in cement increases the amount of entrained air [pp. 293, 295 of 1].
3. An increase in carbon content of fly ash (PFA) and silica fume or bentonite decreases the amount of entrained air [12]. Recent

research suggests that adding polar compounds which absorb into the carbon improves the performance of certain air entraining agents in the presence of pulverized fuel ash [13].

4. The use of water reducing admixtures leads to an increase in the amount of entrained air even if the water reducing admixture has no air-entraining properties per se [pp. 293, 295 of 1, pp. 279, 281 of 7].

5. The more the proportion of fines in aggregate the greater the air content in the concrete [14]. But the material in the 300–600 μm range increases it [pp. 293, 295 of 1].

6. As the maximum size of the coarse aggregate increases, the air requirement of the concrete decreases [pp. 279, 281 of 7]. This effect is indirect, because the larger the maximum size, the smaller the mortar fraction.

7. The amount of air entrained admixture required to obtain a given air content will vary widely depending on the particle shape and grading of the aggregate used [12].

8. More workable mixes holds more air than drier mixes.

9. The higher the slump of the concrete, the higher will be the air content.

10. An increase in water/cement ratio results in an increase in air content and in large air voids [12].

11. Mixing time and speed should be optimized because lower values cause a non-uniform dispersion of the bubbles, while over-mixing gradually expels some air [15, 16].

12. Prolonged transportation and vibration reduce the amount of entrained air (hence, the air content of concrete should be determined just before placement) [pp. 293, 295 of 1].

13. A higher temperature of fresh concrete results in a lower air content, and vice versa. The effect is more significant at higher slumps [pp. 279, 281 of 7].

14. Atmospheric pressure steam curing of concrete may cause expansion of air bubbles and may lead to incipient cracking.

15. Blends of ordinary Portland cement and silica fume in mortars exhibit a higher range of water reducer and air entraining admixture demand [17].

2.5.2 Air void spacing factor and stability

Critical to the durability of concrete undergoing freezing and thawing mechanical actions is the air void spacing factor. The spacing factor and the spacing factor stability can not be determined from the measurement of the air content. The air void spacing factor is measured when concrete is hardened. For mix conditions and materials, the production and the stability of the air void system is related to the air entraining agent dosage.

The real value of laboratory tests to predict field performance of the air-void system of a given mix is questionable and should be reserved for comparative tests. Still more problems arise as a result of the introduction of new materials into the concrete. Some findings are given next.

1. It is possible not only to produce field silica fume concrete with a satisfactory air void spacing factor, but to do so without affecting the stability of the air void system [18].

2. Superplasticizer can sometimes destabilize the air void system significantly. Low dosages of superplasticizers appear to be less harmful that higher ones [19, 20].

3. Retempering concrete with enough water to increase the slump by approximately 50 mm to about 100 mm has no significant influence on the air void spacing factor value although it often causes a small increase in air content [21].

4. Retempering concrete with an air entraining agent diluted in a small amount of water after 45 min, permits the improvement of the spacing factor value if the quantity of admixtures added is high (30 to 50% of the normal dosage) to cause a marked increase of the air content [21].

5. A significant increase in the soluble alkali content of the cement can improve significantly the stability of the air void system, particularly in concrete mixtures to which a superplasticizer is added 15 min after initial mixing [22].

2.6 EFFECT IN MORTAR OR CONCRETE

2.6.1 Fresh stage

(a) Unit mass
Air entrainment affects the unit weight of concrete of mortar.

(b) Workability
The increase in workability brought about by air entrainment is usually ascribed to some sort of ball bearing action of the air bubbles, kept spherical by surface tension [23, 24].

The air bubbles act as fine aggregate of very low surface friction and considerable elasticity, and actually make the mix like an over-sanded mix. For this reason, the addition of entrained air should be accompanied by a reduction in the sand content.

Workability is desirable in all concretes, but those that are inherently harsh and unworkable, such as lean lightweight aggregate concretes, specially benefit from the use of air entrainment.

Consistency An air entrained concrete increases the flowability, at the same water content.

Plasticity   The mix can be said to be more plastic for the same workability. The mix containing entrained air is easier to place than and air free mix.

Cohesion The mix is more cohesive due to the surface tension of the bubbles acting with the cement paste interface, at equal cement content or water/cement ratio [25].

Pumpability High pressure pumping of concretes with a high content of air entrainment causes the bubbles to be compressed, and workability decreases. This is due to the ball bearing effect being diminished when friction increases because of the applied pressure. If the pipe is too long the reduction in air volume due to pressure can absorb all the movement of the piston pump and concrete will not flow through the end. For these reason air entrained concrete is pumped for distances commonly up 45 m [p. 133 of 1].

Compactability Air entrainment increases the compactability because of its ball bearing effect.

Finishability There are thousands of air bubbles per cubic centimetre of cement paste; they allow for easier deformation when the concrete is worked, resulting in an increase of finishability. However, proper magnesium or aluminum alloy floats must be used and a suitable delay may be observed before starting the finishing operations [26, 27]. Indeed, because of the lack of problems caused by bleed water, proper finishing of the air entrained concrete is easy, and a durable surface is ensured.

(c) Segregation
Air entrained concrete is less prone to segregation during handling and transportation but it can not be expected to cure excessive ills that are responsible for segregation, such as excessively lean and wet mixes, poor grading of the aggregate, an improper handling of the concrete [p. 284 of 7].
    Segregation is also improved, providing the fresh concrete is not over-vibrated [p. 298 of 1]. Over-vibration expelled air bubbles.

(d) Freeze-thaw resistance
Air entrainment increases the resistance of concrete to change of volume due to freezing and thawing.

2.6.2 Setting stage

(a) Plastic shrinkage
For the same water/cement ratio and consistency, reduced or unchanged amounts of cement are needed compared to reference mixtures.

(b) Bleeding
The entrained air is beneficial to reduce bleeding. The air bubbles appear to keep the solid particles in suspension so that sedimentation is reduced and water is not expelled [33].

2.6.3 Hardening stage

Strength development   Air entraining admixtures have no appreciable effect on the rate of hydration of cement.

2.6.4 Hardened stage

(a) Unit mass
Air entrainment affects the unit weight of concrete or mortar, inversely to the air content.

(b) Compressive, tensile and flexural strength, modulus of elasticity
Each of these properties is reduced at the same w/c ratio. Some relationships have been found.   The usual approximate rule for predicting the result is that each percent of air will reduce the strength by about 5% [28, 29].

(c) Durability

Capillary absorption   The capillary absorption is lower for properly air entrained mortar or concrete [p. 297 of 7], because of the decrease in water content and because the bubbles reduce the capillary rise.

Permeability to liquids and gases   Air entrainment results in decreasing mixing water for the same workability and results in a decrease in the water/cement ratio; it greatly decreases the permeability of paste [30].

Freeze-thaw resistance   Specimens of cement paste, if kept continually wet, will usually be damaged when frozen, even if they are air-entrained. Depending on the degree of drying, the specimen will be subjected to varied degrees of frost attack, ranging from destruction to apparent immunity. This will depend largely on the properties of the concrete, regardless of the mechanism of frost action.
    Early attempts to explain the damage to concrete caused by freezing were based on the fact that water expands when freezing. Later, however, Collins introduce a concept which was based on frost heaving in soil [31]. This was related to water migrating from unfrozen areas to form ice in large pores, establishing ice lenses and causing

considerable pressure. Powers proposed that the destructive stress is produced by the flow of water away from the region of freezing, and the concrete structure resisting such a flow [32]. Accordingly, if the water content is above the critical saturation point, there will be a critical length of flow path or critical thickness beyond which the hydraulic pressure exceeds the strength of the material, because the resistance to flow is proportional to the length of the flow path. This critical thickness was stated to be of the order of 0.25 mm and accounted for the necessity of entrained air in concrete. The air bubbles were considered to be reservoirs where excess water produced by freezing would migrate without causing pressure.

This hypothesis was modified by Power and Helmuth [33], who concluded that most of the effects of freezing in cement pastes were caused by the movement of water to the freezing sites; the magnitude of pressure generated at a freezing site depended on whether the cavity was filled with solution or ice. In the presence of salt solution, it was considered that pressure may also be generated by osmotic forces due to differences in the concentration of salts in the paste created by freezing of water in large pores.

According to Litvan, some damaged may be caused by the ice formation but the actual process of migration of water is the main source of damage [34]. As this migration is not unlike drying and is initiated only when an ice crystal forms in a large pore (creating a lower vapour pressure), it can be concluded that the latter part of this mechanism does not play a major role. Nevertheless, the importance of water migration should be recognized.

MacInnis and Beaudoin studied the effect of maturity, porosity and degree of saturation on the degree of frost damage in cement paste [35]. They concluded that the major mechanism responsible for frost damage, especially at low levels of maturity, was the hydraulic pressure created in the liquid by the formation of ice. It was suggested that other mechanisms may operate in more mature pastes.

Despite these slightly divergent views, it should be recognized that damage is enhanced by migration of water and that high degrees of saturation and rapid cooling are both detrimental. Air entrainment can be effective in providing reservoirs to prevent the accumulation of ice.

In concrete, the role of aggregates should also be considered. The pores in the aggregates may be such that the pore can be really frozen. Large pores, equivalent to air entrained bubbles (of diameter

600 Å) may not exist in the aggregates. Thus, the volume increase due to freezing of water will either be taken up by the elastic expansion of aggregate of the water flowing out from the aggregate under pressure. According to Powers a maximum of only 0.3% volume of pores can be tolerated in an average aggregate [36]. For saturated aggregates, there must be a critical size [37, 38, 39] below which no frost action occurs, but there is no assurance that the excess water can be accommodated by the air-entrained bubbles in the surrounding paste.

Durability of air entrained slag-cement concretes, with regard to water freezing and thawing, is essentially equivalent to that of concrete containing solely Portland cement. Air entrained slag-cement concretes are somewhat less resistant to laboratory deicer scaling tests than air entrained concrete containing solely Portland cement, despite the fact that both concretes had adequate air-void systems [40].

In general, the addition of silica fume improves both the spacing factor and the specific surface of concrete, whereas tests with high content of silica fume resulted in less favourable values than those shown for the control concrete [41]. The air entrained concrete incorporating 20 and 30% silica fume showed very poor freezing and thawing resistance [42]. It is difficult to entrain more than 5% of air in such concretes and this amount of air may not provide satisfactory void spacing factor values in hardened concrete [43].

The anti-washout concrete cast under water shows poor resistance to freezing-and-thawing compared with the control concrete. The osmotic pressure may be the main reason why the concrete with anti-washout admixtures shows poor durability [44].

It is not easy to predict the probable behaviour of concrete exposed to freezing and thawing. It is generally suggested that potential frost resistance of concrete can only be judged by tests which take into account the environmental conditions to which concrete may be subjected.

Attack by aggressive solutions resistance Air entrainment improves the resistance of concrete to deterioration by sulphate attack [45, 46, 47]. The effect seems to be directly related to the decrease in water/cement ratio for the same workability that is obtained decreasing permeability and to ingress of attacking solutions. Thus, the influence is indirect.

Efflorescence resistance   Efflorescence is lower for properly air entrained mortar or concrete. This reduction occurs because the changes in the capillary structure decrease the migration of salt solution to the surface.

Combined attacks   Malhotra and coworkers [48] in a large scale project related to 9 years field performance in a marine environment and freezing and thawing of normal and lightweight concretes incorporating supplementary cementing materials, reported that the amount of deterioration was bigger with increasing replacement of cement with slag and fly ash.

2.7 Use

To achieve the greatest uniformity in a concrete mixture and in successive batches, it is recommended that air entraining admixtures are added to the mixture in the form of solutions dissolved in the mixing water rather than solids. If other admixtures are also used, the air entraining admixtures should be added separately rather than mixed with the other admixtures, because sometimes there are reactions between materials that result in a decrease in effectiveness of the air entraining agent [p. 274 of 7]. Usually, only small quantities of air entraining admixtures are required to entrain the desired amount of air. These are about 0.005% of active ingredient by mass of cement [p. 292 of 1]. If the admixture is in the form of powder flakes of semisolid, a proper solution must be prepared prior to use, following the recommendation of the manufacturer. If the manufacturers recommended amounts of air entraining admixture do not result in the desired air content, it is necessary to adjust the amount of admixture added.

The alternative way of using air-entraining agents in some countries is to intergrind them with the cement when it is manufactured, obtaining a so-called 'air entraining cement'. The advantage in using air entraining cements is that uncertainties and difficulties that sometimes occur when an admixture is used, can be avoided. The disadvantage is that the air content of concrete is influenced by variables other than the amount of admixture; in that case, one may get a concrete with either less or more air than desired, if an air entraining cement with a fixed amount of admixture is used. Under such circumstances, additional air entraining admixture may be needed.

Attention should be given to proper storage of air entraining admixtures, most of which have a shelf life of at least a year, and are not usually harmed by freezing. The manufacturer's storage recommendations should be followed. Air entraining admixtures are not dangerous, but manufacturer's caution should be followed.

For determining the air content, see RILEM TC 85-TAC Test for Concrete Admixtures.

The mix proportioning of air entrained concrete is not much different to that of non-air entrained concrete, but it must be considered that the air bubbles increase the consistency and workability of concrete and, at the same time, decrease its strength. Because of the change in consistency and workability, less water and fine aggregate are needed for air entrained mixtures, if the consistency is maintained constant. The increase in air volume in the mixture is offset by the decrease in water volume and by the fine aggregate, as when it is designed by the absolute volume method.

2.8 Applications

Because of its greatly improved resistance to frost action, air entrained concrete should be used wherever concrete is exposed to freezing and thawing. The air entrainment increases generally the resistance of concrete to the destructive action of de-icing agents.

Air entraining improves the workability of concrete. It is effective in lean mixtures, which otherwise may be harsh and difficult to work with.   It is beneficial in reducing segregation during handling and transporting, and bleeding.

As a result of its effects in fresh concrete, its use increases the amount of water tightness and, consequently, the durability of concrete.

2.9 Substitute for air entraining agents

Air entraining admixtures are generally used to minimize freezing and thawing attack. Many problems arose trying to adjust the required amount of air with the right amount of bubble size and spacing factor due to the variability of concrete normal ingredients, mixing, retempering, placing methods, supplementary cementing materials, other admixtures and high strength concretes.

Many of the problems involved with the use of air entraining agents were minimized by adding preformed bubbles reservoirs in the form of porous particles.

A method was developed to produce hollow plastic micro-spheres with diameters between 10 and 60 µm. Adding 1% by weight of cement of these microspheres to concrete corresponds to 0.7% by volume of concrete. The spacing factor using these spheres is 0.07 mm which is below the permissible maximum. Adequate frost resistance was attained using 1.0% of the micro-spheres, and in order to be as effective as 5% entrained air, the 28 day strength was higher than the resistance with 5% entrained air [49, 50].

It was demonstrated that porous particles made with fired clay bricks, diatomaceous earth, vermiculite, pumice and perlite can be added to concrete in order to increase its frost resistance. These results show that concretes containing particles are generally more durable than plain concrete [51].

At lower particle concentration the reduction of compressive strength for vermiculite, pumice and perlite has to be lower if compression is made in relation to the air-entrained concrete [52].

2.10 REFERENCES

1. Neville, A.M. (1989) *Concrete Technology*. Longman Scientific & Technical, Harlow.
2. Lawton, E.C. (1939) Durability of concrete pavement: experiences in New York State. *Journal of the American Concrete Institute, Proceedings*, Vol. 35, pp. 561.
3. Moilliet J.L., Collie B. and Blacic, W. *Surface Activity*. Princeton N.J., Van Nostrand and Co.
4. Rosen, M.J. (1978) *Surfactants and Interfacial Phenomena*. New York, John Wiley & Sons, Inc.
5. Schuartz, A.M. and Perry, J.W. (1949) *Surface Active Agents*. New York Interscience.
6. Jackson, F.H. and Timms, A.G. (1954) Evaluation of air entraining

admixtures for concrete. *Public Roads*, 27, pp. 259-67.

7. Dolch, W.L. (1984) Air entraining admixtures. *Concrete Admixtures Handbook: Properties, Science and Technology*, V. S. Ramachandran, editor, Noyes Publications.

8. Klieger, P. (1956) Further studies on the effect of entrained air on strength and durability of concrete with various sizes of aggregates. *Highway Research Board Bulletin* No. 128. Washington, pp. 1-19.

9. American Society for Testing and Materials, Philadelphia (1992). *Recommended Practice for Microscopical Determination of Air Void Content and Parameters of the Air Void System in Hardened Concrete*. ASTM Standard C-457-90.

10. Powers, T.C. (1954) Void spacing as a basis for producing air-entrained concrete. , Vol. 50, No. 9, pp. 741-60, and Discussion December pp. 760-6-15.

11. Backstrom, J.E. Burrows, R.W. Mielenz, R.W., and Wolkodroff, V.E. (1958) Origin, evolution, and effects of the air void system in concrete. Part 2: Influence of type and amount of air entraining agent. *Journal of American Concrete Institute, Proceedings*, Vol. 55. pp. 261.

12. American Concrete Institute Committee 212 (1989) *Guide for Use of Admixtures in Concrete*. pp. 9.

13. Hoarty, J.T. and Hodgkinson, L. (1990) Improved air entraining agents for use in concretes containing pulverized fuel ashes. *Admixtures for Concrete: Improvement of Properties*, E. Vazquez, editor, Chapman & Hall, London, pp. 449-59.

14. Portland Cement Association (1979) *Design and Control for Concrete Mixtures*. 12th Edition, Skokie, Illinois.

15. Neville, A.M. (1970) *Creep of Concrete: Plain, Reinforced and Pre-stressed*, Amsterdam, North Holland Publishing Co., pp. 128.

16. Adams, R.F. and Kennedy, J.C. (1950) *Effect of Batch Size and Different Mixers on the Properties of Air Entrained Concrete*. Laboratory Report No. C-532. US Bureau of Reclamation, Denver, Colorado.

17. Durekovic, A. and Popovic, K. (1990) Superplasticizer and air-entraining agent demand in OPC mortars containing silica fume. *Admixtures for Concrete: Improvement of Properties*, E. Vazquez, editor, Chapman & Hall, London, pp. 1-9.

18. Pigeon, M., Aitcin, P.C. and Laplante, P. (1987) Comparative study of the air-void stability in a normal and a condensed silica

fume field concrete. *ACI Materials Journal*, Vol. 4 No. 3, pp. 194-9.
19. Plante, P., Pigeon, M. and Saucier, F. (1989) Air-void stability, Part 2: Influence of superplasticizers and cement. *ACI Materials Journal*, Vol. 6, pp 581-9.
20. Saucier, F., Pigeon, M. and Plante, P. (1990) Air-void stability, Part 3: Field test on superplasticized concretes. *ACI Materials Journal*, Vol. 7, No. 1, pp. 3-11.
21. Pigeon, M., Saucier, F. and Plante, P. (1990) Air-void stability, Part 4: Retempering. *ACI Materials Journal*, Vol. 7, No. 3, pp. 252-9.
22. Pigeon, M., Plante, P., Plean, R. and Banthia, N. (January-February, 1992) Influence of soluble alkalis on the production and stability of the air-void system in superplasticized and non-superplasticized concrete. *ACI Materials Journal*, Vol. 9, No. 1, pp. 24-31.
23. Mielenz, R.C. (1968) Use of surface active agents in concrete, *Proceedings of Fifth International Symposium on Chemistry of Cement*, Tokyo, Part IV, pp. 1.
24. Bruere, G.M. (1967) Fundamental action of air entraining agents. *RILEM International Symposium on Admixtures*, Brussels III. pp. 5.
25. Powers, T.C. (1944) Mixtures containing intentionally entrained air: topics in concrete technology, Part 3. *Journal PCA Research and Development Laboratories*, Vol. 6, No. 3. Portland Cement Association, Skokie, Illinois, pp. 19-42.
26. Hollon, G.W. and Prior, M.E. (1974) Factors influencing proportioning of air entrained concrete. *ACI Special Publication 46*, American Concrete Institute. Detroit, Michigan, pp. 11.
27. Portland Cement Association. (1960) *Cement Masons Manual.* Skokie, Illinois.
28. Wright, P.G.F. (1953) Entrained air in concrete. *Proceedings, Institution of Civil Engineers*, Part 1, 2, No. 3. London, pp. 337-58.
29. Popovics, S. (1969) Effect of porosity on the strength of concrete. *Journal of Materials*, 4, pp. 356.
30. Powers, T.C. Copeland, L.E., Hayers, J.C. and Mann, H.M. (1954) Permeability of Portland cement paste. *Journal of the American Concrete Institute, Proceedings*, Vol. 51, pp. 285.
31. Collins, A.R. (1944) The destruction of concrete by frost. *Journal Institute of Civil Engineering*. London, pp. 23, 29-41.
32. Powers, T.C. (1945) A working hypothesis for further studies of

frost resistance of concrete. *Journal of the American Concrete Institute, Proceedings*, pp. 245-72.

33. Powers, T.C. and Helmuth, R.A. (1953) Theory of volume changes in hardened Portland cement paste during freezing. *Proceedings of the Highway Research Board*, Vol. 32, pp. 285-97.

34. Litvan, G.G. (1973) Frost action in cement paste. *Materials and Structures*, 6, pp. 293-298.

35. MacInnis, C. and Beaudoin, J.J. (1973) Pore structure and frost durability. *Proceedings of the International Symposium on Pore Structure and Properties of Materials*, Vol. II, F3-F15.

36. Powers, T.C. (November 1965) *The Mechanism of Frost Action in Concrete*. Stanton Walker Lecture Series on the Materials Sciences No. 3. Presented at University of Maryland, National Sand and Gravel Association.

37. Verbeck, G. and Landgren, R. (1960)   Influence of physical characteristics of aggregates on the frost resistance of concrete. *Proceedings ASTM*, 60, pp. 1063-79.

38. Tremper, B. and Spellman, D. (1961)   Tests for freeze-thaw durability of concrete aggregates. *Highway Research Board Bulletin*, No. 305, pp. 28-50.

39. MacInnis, C. and Lau, E.C. (1971) Maximum aggregates size effect on frost resistant of concrete. *Journal of the American Concrete Institute, Proceedings*, Vol. 68, pp. 144-9.

40. Dubovoy, V.S., Gebler, S.H., Klieger, P. and Whiting, D.A. (1986) Effects of ground granulated blast-furnace slag on some properties of pastes, mortars and concretes. *Blended Cements*, ASTM STP 897, G. Frohnsdorff, editor, ASTM, Philadelphia, pp. 29-48.

41. Galeota, D., Giammatted, M.M., Marion, R. and Volta, V. (1991) Freezing and thawing resistance of non air entrained and air entrained concretes containing a high percentage of condensed silica fume. *Durability of Concrete*, ACI SP-126. pp. 249-62.

42. Yamato, T., Emoto, Y. and Soeda, M. (1986)   Strength and freezing-and-thawing resistance of concrete incorporating condensed silica fume. *Fly Ash, Silica Fume, Slag and Natural Pozzolans in Concrete*, V.M. Malhotra, editor, ACI/CANMET, Vol. 2. pp. 1095-1117.

43. Malhotra, V.M.   (1986)   Mechanical properties and freezing-and-thawing resistance of non air entrained and air entrained condensed silica fume concrete using ASTM Test C-666,

Procedures A and B. *Fly Ash, Silica Fume, Slag and Natural Pozzolans in Concrete*, ACI/CANMET, V.M. Malhotra, editor, Vol. 2, pp. 1069-1094.

44. Yamato, T., Emoto, T. and Soeda M. (1991) Freezing and thawing resistance of anti-washout concrete under water. *Durability of Concrete*, V. M. Malhotra, editor, American Concrete Institute SP-126, pp. 169-83.

45. Verbeck, G.J. (1968) Field and laboratory studies of the sulphate resistance of concrete. *Performance of Concrete*, University of Toronto Press, pp. 113-25.

46. Stark, D. (1982) Longtime study of concrete durability in sulphate soils, American Concrete Institute, *ACI Special Publication SP-77*, pp. 21.

47. Bureau of Reclamation, (1950) *Effect of Air Entraining Agent with Sulphate Resistance Cement on the Durability and other Properties of Concrete.* US Bureau of Reclamation, Materials Laboratories Report No. C-362, Denver, Colorado.

48. Malhotra, V.M., Carette, G.G. and Bremner, T.W. (1988) Current status of CANMET's studies on the durability of concrete containing supplementary cementing materials in a marine environment, *Performance of Concrete in a Marine Environment*, American Concrete Institute SP 109-2, V. M. Malhotra, editor. pp. 31-72.

49. Sommer, H. (1978) A new method of making concrete resistant to frost and de-icing salts. *Betonwerk und Fertigteil-Technik*, Vol. 9, pp. 476-84.

50. Vanhanen, A. (1980) *Air entraining agents for frost resistance of concrete.* Laboratory Report No. 73, Technology Research Center, Finland, pp. 73.

51. Litvan, G.G. and Sereda, P.J. (1978) Particulate admixtures and enhanced freeze-thaw resistance of concrete. *Cement and Concrete Research*, Vol. 8, pp. 53-60.

52. Litvan, G.G. (1985) Further studies of particulate admixtures and enhanced freeze-thaw resistance of concrete. *Journal of the American Concrete Institute, Proceedings*, Vol. 82, pp. 724-30.

3

Hardening accelerators

S. Nagataki

3.1 DEFINITION

Hardening accelerators increase the rate of strength gain of concrete at early ages and generally reduce the setting time.

3.2 INTRODUCTION

Hardening accelerators are primarily used in cold weather concreting operations although they may be used in other situations where reduced setting times and early gain of strength are required. Accelerators do not depress the freezing point of water significantly and should not be referred to as 'antifreeze' admixtures.

Hardening accelerators improve early strength, usually because they increase the rate of hydration of C_3S and C_3A, but do not improve long-term strength unless they incorporate water reducers. Most accelerating water reducing admixtures are mixtures of accelerating components blended with water reducing admixtures such as salts of lignosulphonic acids, salts of hydroxycarboxylic acids or low molecular weight polysaccharides [1].

3.3 COMPOSITION

Substances used as accelerators for concrete include alkali hydroxides, silicates, fluorosilicates, calcium nitrite, calcium nitrate, calcium or sodium thiosulphate, calcium or sodium thiocyanate, aluminium chloride, potassium, sodium or lithium carbonate, sodium chloride, calcium chloride and organic compounds such as triethanolamine, formaldehyde and calcium formate [2].

Until recently calcium chloride or admixtures in which calcium chloride is the main active component, have been almost the only accelerators used [3, 18, 20, 21]. Calcium chloride offers many advantages that make it popular as a concrete hardening accelerator. It is very effective in providing a substantial increase in early strength gain and reduces setting time.

In recent years problems have occurred as a result of the corrosion of steel reinforcement promoted by the presence of chloride ions in the concrete. For this reason, several non-chloride accelerators [7] based on calcium formate, calcium nitrite, calcium nitrate, sodium or calcium thiocyanate [3], or triethanolamine have been developed but some of them do promote corrosion. On the other hand, there is one which is a water soluble organic material belonging to the carboxylic acid group [4].

The accelerators are divided into two groups:

- chloride-based accelerators,
- non-chloride accelerators.

3.4 MECHANISM OF ACTION

3.4.1 Chloride-based accelerators

The most common accelerator is calcium chloride. The earliest reference to the use of calcium chloride in concrete is 1885 [6]. Since then it has been extensively used either by itself, or as the main ingredient in many accelerating admixtures.

The accelerating effect of calcium chloride on cement is related mainly to its action on the C_3S phase. Calcium chloride not only alters the rate of hydration of cement minerals but may also combine with them. It also influences such properties as strength, chemical composition, surface area, morphology and pore characteristics of hydration products. Calcium chloride decreases the dormant period in the hydration of C_3S. The increase in strengths at earlier periods with calcium chloride addition may be explained by the increased amount of hydration products formed.

Calcium chloride accelerates the hydration of C_2S. At early periods the accelerating action in C_2S is only marginal compared to that observed in the hydration of C_3S.

Calcium chloride accelerates the reaction between C_3A and gypsum. Calcium chloride reacts with C_3A to form chloroaluminate after gypsum is consumed in the reaction with C_3A.

The effect of calcium chloride on the hydration of C_4AF does not seem to be different from that on C_3A.

A number of investigators [13, 22, 23, 24, 32] have examined its effects on the hydration and hardening behaviour of C_3S. The effects of calcium chloride on the kinetics of the hydration of C_3S, on the morphological features, pore volume and surface area changes, as well as on strength development in the C_3S phase, were studied separately, but no consensus was achieved which would provide a satisfactory explanation for the overall effects of calcium chloride in concrete.

3.4.2 Non-chloride accelerators

The most common accelerators in this class are calcium formate and triethanolamine which are often used to offset the retarding effects of water-reducing admixtures, or to provide non-corrosive accelerators.

Calcium formate accelerates hydration of the C_3S phase of cement. It is, however, not as effective as calcium chloride.

Triethanolamine is known to accelerate the hydration of the C_3A phase of cement and retard the hydration of the C_3S and C_2S phases. It is used as a constituent in admixtures where precise control of the set extension is required. It is also used to offset the retardation effects of other admixtures.

One belonging to the carboxylic acid group is considered to be a catalyst on hydration of the silicates in the cement [4].

3.5 MAIN FACTORS THAT INFLUENCE THE MECHANISM OF ACTION

3.5.1 Type, composition and dosage of admixture

The effect of hardening accelerators depends on the chemical composition and its dosage.

(a) Calcium chloride
Calcium chloride is very effective in providing a substantial increase in early strength and in the rate of setting and reduces both the initial and final setting times of concrete. The optimum dosage of calcium chloride in unreinforced concrete is between 1 and 4%. However, it is generally recommended that the dosage should not exceed 2% by weight of cement, if flaky commercial calcium chloride is used, and 1.5% by weight of cement, if anhydrous calcium chloride is used.

(b) Calcium formate
Calcium formate accelerates setting time and increases the early strength. However, it is not as effective as calcium chloride and a higher dosage is required to impart the same level of acceleration. Calcium formate is sometimes blended with other compounds, for instance sodium nitrite, to obtain enhanced early strength development.

(c) Triethanolamine
Triethanolamine is more effective than calcium chloride in accelerating setting times, consequently only one tenth is required to achieve the same rate of set acceleration as calcium chloride [12]. With up to 0.05%, the initial setting time is retarded slightly but at 0.1% and 0.5%, rapid setting occurs (2, 5, 15].
Triethanolamine reduces strength as the amount is increased [14].

(d) Calcium nitrate
Calcium nitrate accelerates setting times but moderately accelerates hardening.

(e) Calcium nitrite
Calcium nitrite is an effective accelerator for set and hardening.

(f) Sodium thiocyanate
Sodium thiocyanate is a strength accelerator but is not very effective in setting acceleration.

(g) Calcium thiosulphate
Calcium thiosulphate was found to be a better accelerator than the corresponding sodium salt in terms of strength development.

(h) Sodium and potassium carbonate
Sodium and potassium carbonate at dosages greater than 0.1% accelerates setting time [19].

(i) Lithium carbonate
Lithium carbonate accelerates setting time at all concentration.

(j) Carboxylic acid
The carboxylic acid group accelerates setting time and increases the strength.

3.5.2 Type of cement

The accelerating effect depends on the chemical composition if the cement is used as well as the gypsum content [17]. Calcium chloride accelerates slow setting Portland cements more efficiently than faster setting cement. Calcium chloride has an accelerating effect on the hydration of pozzolanic cements. In blast furnace cements, calcium chloride can act as an accelerator at higher temperatures. Under steam curing conditions, it is possible to increase early strength by 15–20% [2].
Concretes containing calcium formate showed accelerated compressive strength when Portland cement containing a low gypsum content was used. Calcium formate is an effective accelerating admixture when the ratio of C_3A to SO_3 is greater than 4 [17].

3.5.3 Temperature

The accelerating effect of calcium chloride is reported to be higher at

0°C and 5°C than at 20°C, and is more pronounced with rich mixtures than with lean ones [9].

3.6 EFFECTS

3.6.1 Fresh stage

(a) Workability
Accelerators have no significant effect on workability. It is however observed that addition of calcium chloride increases slightly the workability and reduces the water required to produce a given slump of concrete (2, 6, 8, 10, 21].

(b) Stiffening
Accelerators shorten the setting times of concrete. Consequently, the consistency loss becomes slightly greater than the reference concrete.

3.6.2 Setting stage

(a) Setting
Accelerators shorten the setting times of concrete. Calcium chloride significantly reduces both the initial and final setting times of concrete.

(b) Heat of hydration
Accelerators increase the rate of hydration of cement and the rate of heat liberated.

(c) Bleeding
Accelerators reduce the rate and amount of the bleeding of the concrete because the hydration reactions during the setting stage are speeded up.

3.6.3 Hardening stage

(a) Heat of hydration
Some reported effects [29] are an increase in the rate of heat evolution

at early stages, but on the calcium chloride total heat of hydration is almost the same in comparison to control concrete [25].

(b) Strength development
The major beneficial effect of the use of accelerator is in the development of high early strength. The maximum rate of increase in strength is reported to be within the first three days of curing [8]. The effect depends on many factors including amount of accelerator added, mixing sequence, temperature, curing conditions, water/cement ratio and cement type.

3.6.4 Hardened stage

(a) Strength
Calcium chloride increases early strength of concrete, but lowers the long term strength. Calcium formate, unlike calcium chloride, increases 28 day compressive strength of concrete [5].

Calcium nitrite increases 1, 3 and 28 day compressive strength.

Sodium thiosulphate accelerates the setting time, but the compressive strengths are slightly reduced with respect to the reference concrete at dosages of 0.5 and 1.0% [5].

Formaldehyde at a rate of 0.07% causes a significant setting time reduction but decreases 28 day compressive strength [5].

(b) Modulus of elasticity
Modulus of elasticity is related to the compressive strength of concrete. Therefore modulus of elasticity is increased at early ages with calcium chloride but at 90 days the values are almost the same for concrete containing 0% or 1–4% calcium chloride of cement.

It has been reported that the modulus of elasticity has not changed after 10 years [26].

(c) Durability

Freeze-thaw resistance   Resistance to freezing and thawing at early ages is increased [25], but at later ages is decreased [2, 8]. To overcome it use of air entraining admixture is recommended.

Attack by aggressive solutions If more than 2% calcium chloride is added, sulphate attack resistance can be lowered [27].

Alkali-aggregate reaction Alkali–aggregate reactivity can be increased by some accelerators [28]. Accelerators containing relatively large amounts of alkali may promote alkali–silica reaction.

Efflorescence Sodium chloride causes efflorescence [28].

Corrosion of reinforcement There has been a great deal of controversy with regard to the use of calcium chloride in reinforced concrete. The presence of chlorides in concrete can promote corrosion activity in the reinforcement. Chloride ions can cause a breakdown in the passive layer that protects the reinforcement against corrosion. In an alkaline medium such as that existing in concrete, corrosion does not occur because the pH value is higher than 10 [2, 7]. But if there are significant amounts of chlorides present, corrosion can occur even when the Ph is in excess of 12.5. The greater the concentration of chloride, the higher the pH value required to protect the reinforcement against corrosion. The actual amount of chloride required to cause depassivation of the reinforcement is not known with certainty.

It is reported that of the other chloride additives stannous chloride is the best. This compound produces the same accelerating effect as calcium chloride and causes less corrosion of reinforcement. It has not been adopted in field practice because of its higher cost [14].

Triethanolamine and calcium formate as non-chloride accelerators are used to provide non-corrosive accelerators.

Calcium nitrite and amino alcohol type accelerators have been shown to improve the corrosion resistance of reinforced concrete [30].

(d) Volume change and creep

Creep The addition of calcium chloride to concrete causes an increase in creep under all conditions [2, 7, 10].

Creep was increased by addition of triethanolamine in concrete specimens loaded after 7 days of curing whereas on difference occurred for specimens loaded at 28 days [16].

Shrinkage   There is some disagreement on the effect of calcium chloride on drying shrinkage. It was found that large amounts of chloride increased the drying shrinkage of concrete. The value decreased as the degree of curing was increased [11].

3.7 USE

3.7.1 Chloride limit

The accelerators containing relatively large amounts of total chloride may accelerate corrosion of steel bar. An addition rate of 2% by weight (77% flake form calcium chloride, by weight of cement) is the most widely used limitation level. But it is not allowed where sulphate resisting cement is used nor in prestressed concrete (CSA, BS, AS).

In West Germany, the content of chloride ion (including other halogen) ions) of admixtures for ordinary concrete and prestressed concrete is limited to be less than 0.2 and 0.1% respectively [31].

Tables 1 and 2 show chloride limits in concrete.

Table 1.   Limits for chloride; ACI Committee 212 IR–81, Chemical Admixtures

Category of concrete	Max. Cl^- ion content by % of cement
Prestressed concrete	0.06
Conventionally reinforced concrete in a moist environment and exposed to chloride.	0.10
Conventionally reinforced concrete in a moist environment but not exposed to chloride.	0.15
Above ground building construction where the concrete will stay dry (does not include locations such as kitchens, parking garages, and waterfront structures, where the concrete will be occasionally wetted).	No limit for construction

Table 2. Limits for chloride; JIS A 5308 Ready Mixed Concrete

	Max. Cl⁻ ion content by kg of 1 m³ concrete
Normal concrete	0.30
When permitted by purchaser	0.60

One of the alternative methods of acceleration when chloride limits must be kept low is the use of chloride-free accelerators.

3.7.2 Proportioning of concrete

Accelerators have no significant effect on the workability or air content, therefore the proportioning of concrete with accelerators is similar to that of the reference concrete.

3.7.3 Dosage

The abnormal setting behaviour by an overdose should be avoided.

The optimum dosage of calcium chloride suggested by various investigators [2, 5, 6, 8] at which the strength is a maximum varies between 1 and 4%. However, it is generally recommended that the dosage should not exceed 2% by weight of cement, if anhydrous calcium chloride is used.

The optimum amount of calcium formate to accelerate the compressive strength appears to be 2 to 3% by weight of cement.

3.7.4 Addition

Accelerators are supplied in solid or liquid form. Calcium chloride is preferred for use in solution form. Calcium chloride should not come directly into contact with cement as it may cause flash set. It is usually added to the mixing water and introduced into the mixer at the same time as other materials.

Where different types of admixture are used, they should be added to the batch separately unless it is known that they can be mixed together satisfactorily. The supplier of the admixtures should recommend proper procedures.

3.7.5 Storage

Different accelerators should be stored in different containers individually. They should be protected against dust or other impurities.

Accelerators in liquid form should be protected against exposure to heat during storage to prevent decomposition and also be protected against freezing. There are admixtures which contain a suspension of solids and the solids may settle with time. These admixtures should be agitated before use.

Accelerators in solid form absorb moisture readily. They should be protected against exposure to moisture to prevent caking or deliquescence.

Accelerators which are stored for a long term and have unusual viscosities, colours or smells, should be retested before use.

3.7.6 Others

In hot weather, some accelerators can produce detrimental effects such as more rapid heat evolution due to hydration, rapid setting and shrinkage cracks. This should be used with care.

The use of calcium chloride for reinforced concrete subjected to steam curing should be avoided because of the corrosion.

3.8 APPLICATION

The benefits of hardening accelerators may include:

* earlier form removal,
* shorter period of protection necessary to avoid damage to concrete by freezing or other factors,

- earlier completion of a structure or repair,
- partial or complete compensation for the effects of low temperatures on rate of strength development.

The benefits of a reduced time of setting may include:

- earlier initiation of surface finishing,
- reduction in pressure on forms or reduction of length of time during which forms are subjected to hydraulic pressures,
- more effective plugging of leaks against hydraulic pressure.

3.9 SPECIFICATIONS

Accelerators are recognized by American, European, Australian and other countries. The following are some standards.

Standards	*Type of admixture*
ASTM C-494 (USA)	Type C; accelerating
BS 5075 (UK)	accelerating
AS 1478 (Australia)	Type AC; set-accelerating
AFNOR P18-103 (France)	accelerating
JIS A 6204 (Japan)	

3.10   REFERENCES

1. Concrete Society. (1980) *Guide to Chemical Admixtures for Concrete*, Technical Report No 18, London.
2. Ramachandran, V. S., (1984) *Concrete Admixture Handbook: Properties, Science and Technology*, Noyes Publications, USA.
3. Non-chloride accelerating admixtures, *Concrete Construction*, April (1985).
4. Popovics, S., (1990) A study on the use of a chloride-free accelerator, *Admixtures for Concrete: Improvement of Properties*, E. Vazquez, editor, Chapman & Hall, London, pp. 197–208.
5. Rosskopf, P. A., Linton, F. J. and Peppier, R. B., (1975) Effect of

various accelerating chemical admixtures on setting and strength development of concrete, *Journal of Testing and Evaluation*, Vol. 3, p. 322–30.

6. Ranga Rao, M. V., (1976) Investigation of admixtures for high early strength in concrete, *Indian Concrete Journal*, pp. 279.
7. Calcium chloride in concrete, (1976) *Indian Concrete Journal*.
8. Erlin, B. and Hime, W. G., (1976) The role of calcium chloride in concrete, *Concrete Construction*.
9. Warris, B., (1968) Effect on admixtures on the properties of fresh mortar and concrete, *Materials and Structures*, p. 97, March–April.
10. American Concrete Institute, (1984) *ACI Manual of Concrete Practice*, Part 1.
11. Bruere, G. M., Newbegin, J. D. and Wilson, L. M., (1971) *A Laboratory Investigation of the Drying Shrinkage of Concrete Containing Various Types of Chemical Admixtures*, Technical Paper No. 1, Division of Applied Mineralogy, CSIRO, Australia.
12. Collis, M., (1982) Accelerating and retarding admixtures, *Seminar on Admixtures*, Parkville, Concrete Institute of Australia.
13. Skalny, J. and Odler, I., (1967) The effect of chlorides upon the hydration of Portland cement and upon same clinker hydration of Portland cement and upon same clinker minerals, *Magazine of Concrete Research*, Vol. 19, p. 203.
14. Ramachandran, V. S., (1973) Action of triethanolamine on the hydration of tricalcium aluminate, *Cement and Concrete Research*, Vol. 3, No. 1, pp. 41–54.
15. Bruere, G. M., (1982) Admixtures chemistry, *Seminar on Admixtures*, Parkville, Concrete Institute of Australia.
16. Hope, B. B. and Manning, D. G., (1971) Creep of concrete influenced by accelerators, *Journal of the American Concrete Institute, Proceedings*, Vol. 68, No. 5, p. 361-5, May.
17. Gebler, S., (1983) Evaluation of calcium formate and sodium formate as accelerating admixtures for Portland cement concrete, *Journal of the American Concrete Institute, Proceedings*, Vol. 80, No. 5, p. 439–44, September–October.
18. New Zealand Concrete Research Association, (1982) *Admixtures for Concrete*, Information Bulletin, IBO13.
19. Valenti, G. L. and Sabatelli, V., (1980) The influence of alkali carbonates on the setting and hardening of Portland pozzolanic cements, *Silicates Industriels*, Vol. 45, p. 237.

20. Stallworthy, R. A., (1982) Accelerating and retarding admixtures. Problems and effect in concrete, *Seminar on Admixtures*, Parkville, Concrete Institute of Australia.
21. Department of Housing and Construction, Australia, (1981) *Admixtures for Concrete 1 and 2*, Notes on the Science of Building, NSB 100A and NSB 100B.
22. Ramachandran, V. S., (1971) Kinetics of tricalcium silicate in presence of calcium chloride by thermal methods, *Thermochimica Acta*, Vol. 2, p. 41.
23. Odler, L. and Skalny, J., (1971) Influence of calcium chloride in paste hydration of tricalcium silicate, *Journal of the American Ceramic Society*, Vol. 54, p. 362.
24. Ramachandran, V. S., (1972) Elucidation of the role of chemical admixtures in hydrating cements by DTA technique, *Thermochimica Acta*, Vol. 3, pp. 343–66.
25. Kobayashi, M. (1976) Admixtures for concrete (in Japanese), *Cement Concrete*, No 354, p. 50, August.
26. Sone, T. and Ooshio, A. (1984) A long term performance test of high strength concrete using hardening accelerator (in Japanese), *Proceedings of Annual Meeting of Japan Society of Civil Engineers*, Vol. 39, V–66, October.
27. Kasai, Y. and Kobayashi, M., (1986) Admixture for cement and concrete, *Gijutsushoin*, p 337.
28. Ooshio, A., (1970) Retardation, high-early-strength, quick-hardening, bleeding etc. (in Japanese), *Concrete Journal*, Vol. 8, No. 3, p. 45.
29. Kantro, D. L., (1975) Tricalcium silicate hydration in the presence of various salts, *Journal of Testing and Evaluation*, Vol. 3, No. 4, p. 312-32.
30. Kuroda, T., Goto, T. and Kobayashi, S., (1986) Non-chloride and non-alkali metal hardening accelerator (in Japanese), *Cement Gijutsu Nempo*, Vol. 40, p. 226.
31. Richtlinien fur die Zuteilung von Prufzeichen fur Betonzusatzmittel (1973).
32. Abdelrazig, B. E. I. *et al*, (1990) Effects of accelerating admixtures on cement hydration, *Admixtures for Concrete: Improvement of Properties*, E. Vazquez, editor, Chapman & Hall, London, pp. 106–19.

4

High-range water reducers (superplasticizers)

S. Biagini

4.1 DEFINITION

High-range water reducing agents (superplasticizers) are admixtures which procure a considerable increase in the workability of mortars and concretes at constant water/cement ratio. The duration of the effect is generally temporary and variable. Mortars and concretes of constant workability can be made with smaller amounts of water, saving more than 12% without undue retardation, excessive entrainment of air or detrimental bleeding.

4.2 INTRODUCTION

'High-range water reducers (superplasticizers)' are subjected in different countries to a control of their ability to reduce mixing water or increase the workability of cement mixes. The limit of the improvement has been quantified in order to distinguish these admixtures from those defined as 'water reducers'.

4.3 COMPOSITION

The high-range water reducers presently used in the market can be classified [1] according to their chemical nature in the following main groups:

- beta-naphthalene sulphonate formaldehyde condensates;
- melamine sulphonate formaldehyde condensates;
- modified lignosulphonates;
- esters of sulphonic acids;
- salts of carboxylic/hydroxycarboxylic acids.

There is also a large number of patents [2, 3] claiming the fluidity ability of a series of different chemical compounds, but the commercial importance of these admixtures is today negligible.

4.4 MECHANISM OF ACTION

The mechanism of action of high-range water reducers is mainly based on their ability to be adsorbed on the surface of cement particles and modify the rheological behaviour of the cement matrix [4, 5]. The rate of adsorption of high-range water reducers depends on the chemical and mineralogical composition of the cement, its fineness and in particular on the C_3A ($C_3A = 3CaO.Al_2O_3$) content. It has been found that calcium aluminate [6–9] adsorbs very rapidly the high-range water reducer molecules, while calcium silicate in the first hours of hydration adsorbs only a lower amount of the high-range water reducers.

The increase of workability that can be obtained in a concrete by the use of superplasticizers can be correlated with the following properties.

1. The value of zeta potential of the electric double layer that is formed on the surface of the cement particles by the polar groups of adsorbed superplasticizer chains [10–14].
2. The molecular weight of the superplasticizer [15, 16].

The rate of workability loss is correlated to the retardation produced on the hydration of cement.

4.5 EFFECTS

The different properties of concrete are examined and the alternative names high-range water reducers or superplasticizer will indicate the specific way of using the admixture, that produces the modification described. It is intended that if one of the two names of the admixture is not indicated, the corresponding way of using the admixture does not exert a particular action on the specific property cited.

4.5.1 Fresh state

(a) Unit mass
Unit mass of concrete is usually increased when high-range water reducers are used.

(b) Workability

Consistency   Superplasticizers dramatically increase the ability of concrete to flow.

Cohesion   Cohesion is largely improved by the use of high-range water reducers as a consequence of the reduction of water in concretes.

Air content   Air content may be slightly increased, especially in the case of use of high dosages of the admixtures as superplasticizer.

Slump loss   At the same initial workability, slump loss may be higher in concretes with high-range water reducers than in concrete without admixture. At the same water/cement ratio, slump loss of concrete with superplasticizer may be higher or lower than the control concrete without admixture as a function of the type of superplasticizer used.

Pumpability   Pumpability of concrete is improved by the use of superplasticizer, as a consequence of the increase in workability, and due to cohesion in case of use as high-range water reducers.

(c) Segregation
Segregation decreases when the admixture is either used as a high-range water reducer or as a superplasticizer, provided that an adequate mix design of the concrete is done.

4.5.2 Setting state

(a) Setting
Generally the admixture used as a superplasticizer mildly retards the setting of concrete [23], while use as a high-range water reducer at normal dosage does not give significant retardation.

(b) Plastic shrinkage
Plastic shrinkage cracking can be increased by the use of high-range water reducers if the ambient conditions are such that evaporative demands are greater than the reduced bleeding capacity of the high-range water reduced concrete.

(c) Bleeding
Bleeding is reduced by the use of high-range water reducers. If the aggregate size distribution is not properly designed, bleeding can be increased when superplasticizer is used [1].

4.5.3 Hardened state

(a) Strength
The strength of concretes is considerably increased by the use of high-range water reducers as a consequence of the reduction of the water/cement ratio [24], while strength is not substantially modified in case of use as a superplasticizer.

(b) Porosity

Capillary absorption   Capillary absorption of concrete is strongly reduced when the admixture is used as a high-range water reducer.

Permeability   The permeability of concrete is directly linked to its capillary porosity which is influenced by the water/cement ratio, that can be largely reduced by the use of the admixture as a high-range water reducer.

Freeze-thaw attack   High-range water reducers-superplasticizers normally induce some air entrainment in the concrete mixes, but some of the air bubbles introduced are larger than those of air entraining agents, and therefore are not useful to increase the freeze-thaw durability of concretes [26].

Attack by aggressive solutions   The resistance of concretes to attack by aggressive solutions is increased by high-range water reducers because of the reduction of concrete capillarity porosity. The use of the admixture as superplasticizer does not change the resistance of the concrete [26, 27, 28].

(c) Volume change

Creep   The use of high-range water reducers reduces creep due to the reduction of the water/cement ratio of the concrete.

Drying shrinkage   The shrinkage of concrete is reduced by high-range water reducers mainly because of the reduction of the water content of the concrete. When a concrete is manufactured with the admixture used as a superplasticizer, its shrinkage, for the same percentage of moisture loss, has been found [27] to be higher than in a concrete produced with the same quantity of water but without the use of the superplasticizer. On the other hand it has been also shown [19] that with the same curing conditions, the shrinkage of a superplasticized concrete is similar to that of a corresponding plain concrete.

The conclusion can be drawn that the better dispersion of cement particles in a superplasticized concrete produces a finer capillary structure, which reduces the rate of moisture loss of the concrete under normal ambient conditions, so that the shrinkage of superplasticized concrete is practically similar to that of a normal concrete manufactured with the same amount of water.

4.6 USE

The admixtures classified 'high-range water reducers/superplasticizers' are normally aqueous solutions, with a dry active material content between 30 and 40% by weight. They are commonly added to the cement mix together with the mixing water. However, addition after the mixing water has been found to be beneficial for obtaining better performance.

There are some examples of superplasticizer used in powder form, and in this case they are added to the aggregates and the cement before the mixing water is added.

The optimum dosage of high-range water reducers can vary from one product to other, and must be used according to the producer recommendation.

4.7 APPLICATIONS

The success of high-range water reducers in the market is due to a series of advantages that can be summarized as follows:

(a) Use as high-range water reducers
Reduction of the water/cement ratio with a consequent improvement of all characteristics of the concrete (increase of compressive and flexural strength, decrease of permeability, creep, shrinkage).

(b) Use as superplasticizer
Production of high quality concretes with high workability [17–20]. Easy placing of concretes also in case of presence of high density of steel reinforcement.

4.8 REFERENCES

1. Ramachandran, V. S., (1984) *Concrete Admixtures Handbook: Properties, Science and Technology*, Noyes Publications, NJ, USA.
2. Rixom, M. R. and Mailvaganam, N. P., (1978) *Chemical Admixtures for Concrete*, E & FN Spon, London.
3. American Ceramic Society, Columbus, Ohio, USA, *Cement Research Progress*, Volumes from 1974.
4. Kondo, R., (1976) *Reaction of Cement and Cement Hydrate with Organic Compounds*, Report to the Ministry of Education, pp. 3, 35.
5. Collepardi, M., Corradi, M. and Valente, M., (1980) Influence of polymerization of sulfonated naphthalene on the fluidity of cement paste, *Proceedings of Seventh International Congress on Chemistry of*

Cements, Paris, Vol III, pp. 20-5.

6. Ramachandran, V. S., (1983) Adsorption and hydration behaviour of tricalcium aluminate-water and tricalcium aluminate-gypsum-water systems in the presence of super-plasticizers, *Journal of the American Concrete Institute, Procedings*, Vol. 80, pp. 235-41.

7. Ramachandran, V. S., (1981) Influence of superplasticizers on the hydration of cement, *Proceedings of Third International Congress on Polymers in Concrete*, Koriyama, Japan, pp. 1071-81.

8. Ramachandran, V. S., (1972) Elucidation of the role of chemical admixtures in hydrating cements by DTA technique, *Thermochimica Acta*, Vol. 3, pp. 343-66.

9. Ramachandran, V. S., (1972) Effect of calcium lignosulfonate on tricalcium aluminate and its hydration products, *Materiaux et Constructions/Materials and Structures*, Vol. 5, pp. 67-76.

10. Collepardi, M., Corradi, M. and Valente M., (1981) Influence of polymerization of sulfonated naphthalene condensate and its interaction with cement, *Developments in the Use of Superplasticisers*, American Concrete Institute, ACI SP-68, pp. 485-98.

11. Petrie, E. M., (1976) Effect of surfactant on the viscosity of Portland cement water dispersions, *Industrial Engineering Chemistry*, Vol. 15, pp. 242-9.

12. Roy, D. M. and Daimon, M., (1980) Effects of admixtures upon electrokinetics phenomena during hydration of C_3S, C_3A and Portland cement, *Proceedings of Seventh International Congress on Chemistry of Cements*, Paris, Vol. II, pp. 242-6.

13. Ernsberg, F. M. and France, W. G., (1945) Portland cement dispersion by adsorption of calcium lignosulfonate, *Industrial Engineering Chemistry*, Vol. 37, pp. 598-600.

14. Blank, B., Rossington, D. R. and Weinland, L. A., (1963) Adsorption of admixtures on Portland cement, *Journal of the American Ceramic Society*, Vol. 46, pp. 395-9.

15. Basile, F., Biagini, S., Ferrari, G. and Collepardi, M., (1986) Effect of condensation degree of polymers based on naphthalene on fluidities of cement pastes, *Proceedings of Eighth International Congress on the Chemistry of Cement*, Rio de Janeiro, Vol. VI, pp. 260-3.

16. Basile, F., Biagini, S., Ferrari, G. and Collepardi, M., (1986)

Properties of cement mixes containing naphthalene sulfonated polymers of different molecular weight, *Proceedings of the Eighth International Congress on Chemistry of Cement*, Rio de Janeiro, Vol. VI, pp. 264–8.

17. Massazza, F., Costa, U. and Corbella, E., (1977) Influence of beta-naphthalene sulphonate formaldehyde condensate superplasticizing admixture on C_3A hydration, *Seminar on Reaction of Aluminates during the Setting of Cements*, Eindhoven, The Netherlands.

18. Ramachandran, V. S., (1980) Hydration of C_3A in the presence of admixtures, *Proceedings of the Seventh International Congress on Chemistry of Cements*, Paris, Vol. IV, pp. 520–3.

19. Collepardi, M. and Corradi, M., (1977) High strength and reliable concretes, *International Conference on Cement and Concrete Admixtures and Improving Additives*, Mons, Belgium; also *Silicates Industriels*, pp. 44, 13 (1979).

20. Collepardi, M. and Massidda, L., (1976) *Hydraulic Cement Pastes: Their Structure and Properties*, Proceedings of Conference, Sheffield, UK, Cement and Concrete Association, p. 256.

21. Sinka, J., Fleming, J. and Villa, J., (1978) Condensates on the properties of concrete, Presentation at *First International Symposium on Superplasticizers*, Ottawa, Canada, p. 22.

22. Bocca, P., (1978) *Modulo*, pp. 3, 267.

23. Malhotra, V. M., (1979) *Superplasticizers: Their Effect on Fresh and Hardened Concrete*, CANMET, Canada, p. 23.

24. Aignesberger, A. and Kern, A., (1981) Use of melamine-based superplasticizer as a water reducer, *Developments in the Use of Superplasticisers*, American Concrete Institute, SP-68, pp. 61–80.

25. Collepardi, M. and Corradi, M., (1978) Influence of naphthalene-sulfonated polymer based superplasticizers on the strength of lightweight and ordinary concretes, *Superplasticisers in Concrete*, V. M. Malhotra, editor, American Concrete Institute, ACI SP-62, Vol. 2, pp. 451–80.

26. Mielenz, R. C. and Sprouse, J. H., (1978) *Superplasticisers in Concrete*, V. M. Malhotra, editor, American Concrete Institute, ACI SP-62, Vol. I, p. 1.

27. Brooks, J. J., Wainwright, P. J. and Neville, A. M., (1978) Time-dependent properties of concrete containing Mighty admixtures, *Superplasticisers in Concrete*, V. M. Malhotra, editor, American Concrete Institute, ACI, SP-62, Vol. 2, pp. 425–50.

28. Mukherijee, P. K. and Chojnacki, B., (1978) Laboratory evaluation of a concrete superplasticizing admixture, *Superplasticisers in Concrete*, V. M. Malhotra, editor, American Concrete Institute, ACI SP-62, Vol. 1, pp. 403-24.
29. Ghosh, R. S. and Malhotra, V. M., (1978) *Use of Superplasticizers as Water Reducers*, CANMET Division Report MRP/MRL 78-189(J), Ottawa.
30. Kishitani, K., Kasami, H., Iizuka, M., Ikeda, T., Kazama, Y. and Hattori, K., (1981) Engineering properties of superplasticized concretes, *Developments in the Use of Superplasticisers*, American Concrete Institute, ACI SP-68, pp. 233-52.

5

Water reducing/retarding admixtures

M. Ben-Bassat

5.1 DEFINITION

(a) Water reducer
A water reducing admixture can be defined as an admixture that reduces the amount of the mixing water of mortar and concrete for a given workability. They can sometimes have a secondary effect of retardation or acceleration on the setting time of the concrete.

(b) Retarder
A retarding admixture can be defined as an admixture which retards the set and the initial hardening of the concrete.

5.2 INTRODUCTION

This chapter summarizes present knowledge of the use of the water reducing/ retarding chemical admixtures in concrete.

In general, water reducers/retarders reduce the water requirement of the mix and, at the same time, can modify the properties of the fresh and hardened concrete. The admixtures are classified according to type of constituting materials, or the characteristic effect of their use. Commercial admixtures may contain materials that separately would belong in two or more groups.

Water reducing/retarding admixtures may be used to modify the properties of the fresh and hardened concrete in such a way as to make it more suitable for work in modern concrete technology, and gain desired concrete properties.

The successful use of admixtures depends upon proper selection and dosage.

5.3 WATER REDUCERS

5.3.1 Composition

The main materials used in water reducers are:

1. lignosulphonic acids and their salts;
2. modifications and derivatives of lignosulphonic acids and their salts;
3. hydroxylated carboxylic acids and their salts;
4. modifications and derivatives of hydroxylated carboxylic acid and their salts;
5. other materials which include: zinc salts, phosphates, chlorides, carbohydrates, polysaccharides, and sugar acids. Certain polymeric compounds, melamine derivatives, naphthalene derivates, and others.

5.3.2 Mechanism of action

Water reducing admixtures can be divided into three groups:

- normal water reducers
- retarding water reducers
- accelerating water reducers.

Water-cement systems usually exhibit flocculations of solid particles which tend to agglomerate into clusters. In the presence of a water reducing admixture a reduction of the attraction forces between particles occurs and, as a result, the clusters break down and the particles become more mobile. Some mechanisms explain the reduction of these attraction forces when water reducers are added to a given concrete mix by:

- reduction of interfacial tension [1]
- multilayer absorption of organic molecules [2, 32]
- increase in electrokinetic potential [16, 23]
- protective adherent sheet of water molecules [1]
- release of water trapped in cement clumps [4]
- change in the morphology of the hydrated cement compounds [15, 18].

5.3.3 Factors influencing mechanisms of action

(a) Composition, brand and dosage of admixture
The effect of different brands or types of commercial water reducers on water reduction depends on the chemical composition and its concentration [23, 30]. The higher the dosage of the admixture the higher the water reduction. Generally increasing content, for example more than 0.1% (solid content of admixture by weight of cement), does not yield much additional effect in reducing the water content.

(b) Type, brand and content of cement
Cement composition, especially C_3A/C_3S ratio and C_3A content, as well as a sulphate form, may affect the efficiency of the admixture [11, 23, 24].

(c) Type of aggregate
In some cases aggregates from a general source may have a different effect on water reduction compared to those from another source which differ in grading, shape, physical properties, mineralogical composition, etc [30].

(d) Types and amount of additive such as slags and fly ash
A higher dosage of admixture is usually required in concretes containing slag or fly ash in order to obtain the same water reduction as that of mixes without slags or fly ash. This could be due to higher absorption of water reducers by the additive. In the case of silica fume, high-range water reducers are required.

(e) Temperature
Except for setting time temperature does not affect dosage.

5.3.4 Effects on fresh concrete

(a) Unit mass
For a given workability the concrete unit mass can be increased by adding water reducing admixtures [30].

(b) Workability
Workability of concrete is improved when water reducers are incorporated into the mix at a given water content. For a given consistency, the admixture-treated concrete with reduced water generally is more workable [7, 10]. Improving concrete workability by admixture treatment means better appearance [6, 10], uniformity [10, 29], pumpability [31] and finishing characteristics [6, 31].

It has been found [6, 23] that the admixture efficiency and capability to reduce water content is higher for high slump concrete.

Concretes containing admixtures are more sensitive to retempering. The additional amount of water needed to reach a required variation in slump would be smaller for concrete with admixture than the amount of water needed for concrete without admixture [7, 10].

(c) Bleeding
For a given slump water reducers can modify the rate or capacity for bleeding or both.

(d) Heat of hydration and temperature rise
When water reducers are used, then heat of hydration and temperature rise of concretes may be higher at early ages. When the water reducer is used to reduce the mixing water of concrete for a given workability, without modifying the mix proportions.

(e) Plastic shrinkage
Generally, plastic shrinkage occurs by loss of moisture during the first 2 to 5 h from placing and compacting of the concrete. If plastic shrinkage causes cracking, water reducing admixtures can aggravate the cracking problem.

5.3.5 Effects on hardened concrete

The effect of water reducing admixtures on the properties of hardened concrete depends mainly on the mode of use of these admixtures.

(a) When the admixture is used to improve workability without modifying the mix proportions, there is an insignificant influence on the properties of the hardened concrete.

(b) Reduction of both cement and water by incorporating admixtures to maintain workability level and the designed strength also have an insignificant influence on the properties of the hardened concrete.

(c) Reducing the mixing water of the concrete mix by water reducer, in order to maintain the workability level without modifying the cement content, will have a substantial influence on the properties of the hardened concrete. The physical properties (specific gravity, porosity, surface properties, permeability, drying shrinkage, creep, etc), mechanical properties (compressive strength, flexural strength, etc) and chemical properties (durability, sulphate resistance, etc) are affected by the water reducers, particularly when the admixtures are used to reduce water, to reduce the water/cement ratio and to perform a better concrete class. There will be no further discussion on this issue in this chapter.

5.3.6 Use

Water reducers/retarders should be used according to the dosage recommended by the supplier. Typical recommended dosages of the admixture are in the range of 0.2–0.4% of the cement content, by mass.

Admixtures of all classes are available in either powder or liquid form. Since relatively small quantities are used, it is important that suitable and accurate dispensing equipment should be used. For water reducers/retarders it is preferable to use the admixture in liquid batching system.

(a) Admixture addition procedure [23]
Admixtures should be incorporated in the concrete mix in such a way that the most rapid and uniform dispersion of the admixture is obtained. Maximum benefit is obtained by adding the admixture just at the end of the mixing period of the aggregates, cement and water. However, with this procedure there are some practical difficulties in obtaining the desired workability and the uniform dispersion of the admixture, particularly when large concrete mix batches have to be prepared, like in a ready-mixed concrete operation.

A reasonable compromise between good technical performance and practical use of water reducers can be achieved by the following procedure.   After an initial mixing period of 15–30 seconds of aggregates, cement, and approximately 50% of the mixing water, the admixture dissolved in part (up to 30%) of the gauging water is introduced into the mix, and finally the remainder of the water is added, until the required workability is obtained.   The addition sequence can also affect the setting time and the compressive strength. Influence of the addition procedure of the admixture can also be assessed in terms of different dosage used to achieve approximately the same concrete performance.

(b) Testing
A test should be made to evaluate the effect of the admixture on the properties of concrete made with job materials under the anticipated ambient conditions and construction procedures to be used. Tests of water reducing admixtures and set controlling admixtures should indicate their effect on the following properties of the concretes:

* water requirement
* air content
* consistency
* bleeding of water and possible loss of air from the fresh concrete
* time of setting
* compressive (flexural) strength
* resistance to freezing and thawing
* drying shrinkage.

All the properties should be tested according to the test procedures as specified in the relevant standards or codes of practice.

Accidental overdosage of 2-3 times the supplier's recommended dosage, can cause a significant retardation in setting time and a decrease in early compressive strength (24-48 hours), increase of air entrained and changes of other properties. It has been found [31] that in these cases, extended curing does not damage compressive strength at later ages.

Before use, admixtures should be tested for one or more of the following reasons:

1. to determine the compliance with a purchase specification;
2. to evaluate the effect of the admixtures on the properties of the concrete;
3. to determine within batch uniformity of the product.

During the use of the admixture, tests should be done periodically to provide data showing that any batch is the same as that previously supplied.

5.3.7 Application

There are three major modes of use of these admixtures:

1. to reduce the content of mixing water of concrete for a given workability;
2. to reduce both water and cement content so that workability and strength water/cement ratio of the concrete are similar to those of the control concrete mix;
3. to improve workability without modifying the mix proportions.

The decision to use the admixture in one of the modes described above depends on the comparative gain in cement saving, strength increase, or improved workability, to attain the designed performance properties of the concrete mix.

Some of the more important engineering purposes for which admixtures are used are listed here:

• to increase workability without increasing water content or to decrease the water content at the same workability;

- to reduce or prevent settlement;
- to modify the rate or capacity for bleeding, or both;
- to reduce segregation;
- to improve pumpability;
- to accelerate the rate of strength development;
- to increase durability;
- to decrease permeability;
- to increase bond of concrete to steel reinforcement.

5.3.8 Standards and codes of practice

Different types of water reducers, water reducers/retarders and retarders are recognized by the Standards in Europe, Japan, Canada, America and other countries. In Tables 1–3 [23] a comparison is given of standard requirements in different countries.

ACI committee report 212 [1] gives a wide review and a guide for use of admixtures in concrete.

5.4 RETARDERS

5.4.1 Composition

The ingredients used as retarders are very much the same as the ingredients used for water reducers (see Section 5.3.1). The degree of effect depends upon the relative amounts of each ingredient used in the formulation, type, brand and content of cement [4, 5, 15, 19].

5.4.2 Mechanism of action

Retarding admixtures can be divided into two groups:

- normal retarder,
- retarder/water reducer.

When a retarder is added to the cement–water system, physical adsorption and chemical reactions occur generally with the cement

Table 1: Comparison of Standard Requirements for Water-Reducing Admixtures (23)

Standard	Type of admixture	Initial setting time (with respect to control mix)	Water reduction, min. (with respect to control mix)	Compressive strength, min. (% of control mix)	Shrinkage max. (% of the control mix)
ASTM C – 494 – 81	Type A: water reducing	Min: 1 h earlier Max: 1 1/2 h later	5%	110 (3, 7, 28 days)	135 *
BS 5075: 1982 [x]	Normal water reducing	Min: 1 h earlier ** Max: 1 h later **	8%	110 (7, 28 days)	–
CAN 3 – A266 2 – M78	Type WN: normal setting	Min: 1h 20 min earlier Max: 1h 20 min later	5%	115 (3, 7, 28 days)	135 *
AFNOR NF P 18 – 103	Water reducing •	Min: 1 1/2 h earlier Max: 3 h later	6.5 ***	110 (3, 7, 28, 90 days)	105
JIS A 6204	Type A: water reducing	Min: 1 h earlier Max: 1 1/2 h later	4%	115 (3 days) 110 (7 and 28 days)	120
UNI 7102 – 72	Water reducer (plasticizer)	After 1 h on cement paste	5%	105 (1, 3 days) 110 (7 days) 115 (28, 90 days)	0.01 [+]

* % of control limit applies when shrinkage of control is not less than 0.030%; when shrinkage of control is less than 0.030%, the increase over control limit should be less than 0.010%.

** Penetration resistance 0.5 N/mm².

*** A minimum value of 8.6% is required for low – C_3A cements (≤ 4%).

+ Maximum increase over control (mortar).

x Compacting factor relative to control mix: not more than 0.02 below.

• At the same workability of the control mix.

Table 2: Comparison of Standard Requirements for Water-Reducing and Retarding Admixtures (23)

Standard	Type of admixture	Initial setting time (with respect to control mix)	Water reduction, min. (with respect to control mix)	Compressive strength, min. (% of control mix)	Shrinkage max. (% of the control mix)
ASTM C – 494 –77	Type D: water reducing and retarding	Min: 1 h later Max: 1 1/2 h later	5%	110 (3, 7, 28 days)	135*
CAN 3 A266.2 – M78	Type WR: water reducer – set retarder	Min: 1 h later Max: 3 h later **	5%	115 (3, 7, 28 days)	135*
BS 5075 – 82 [x] 2 – M78	Retarding – water reducing	Min: 1 h later **	8%	110 (7, 28 days)	135*
JIS A 6204	Type D: water reducing and retarding	Min: 1 h later Max: 3 h later	5%	105 (3 days) 110 (7, 28 days)	120
UNI 7107 – 72	Retarding water reducer	Min: 3/4 h later	5%	110 (3 days) 110 (7 days) 115 (28 days)	0.01+

* % of control limit applies when shrinkage of control is not less than 0.030%; when shrinkage of control is less than 0.030%, the increase over control limit should be less than 0.010%.

** Penetration resistance of 0.5 N/mm².

+ Maximum increase over control (mortar).

x Compacting factor relative to control mix: not more than 0.02 below.

Table 3: Comparison of Standard Requirements for Retarding Admixtures (23)

Standard	Type of admixture	Initial setting time (with respect to control mix)	Water reduction, min. (with respect to control mix)	Compressive strength, min. (% of control mix)	Shrinkage max. (% of the control mix)
ASTM C – 494 – 81	Type B: regarding admixture	Min: 1 h earlier Max: 3 1/2 h later	5%	90 (3, 7, 28 days)	135*
BS 5075 – 82 [+]	Retarding	Min: 1 h later**	5%	90 (7 days) 95 (28 days)	135*
CAN 3 – A266.2 – M78	Type R: moderate set retarder	Min: 1 h later Max: 3 h later	3%	110 (3, 7, 28 days)	135*
CAN 3 – A266.2 –78	Type R_x: extended set regarder	Min: 5 h later	3%	100 (3 days) 110 (7, 28 days)	135*
AFNOR NF P 18 – 103	Retarding• admixture	Min: 3/4 h later to 24 h Range initial and final setting time ≤ 8 h		80 (7 days) 90 (28 days) 100 (90 days)	125 (7 days) 110 (28, 90 days)
UNI 7104 – 72	Retarding admixture	Min: 3/4 h later		100 (28 days)	0.01 [+]

* % of control limit applies when shrinkage of control is not less than 0.030%; when shrinkage of control is less than 0.030%, the increase over control limit should be less than 0.010%.

** Penetration resistance 0.5 N/mm².

+ Maximum increase over control (mortar).

x Compacing factor relative to control mix: not more than 0.02 below that of the control mix.

• At the same workability of the control mix.

components, and especially with the C_3A and the C_3S. The retardation of the setting times of the concrete is very sensitive to the generic group of the retarder and the cement composition.

5.4.3 Factors influencing mechanisms of action

See Section 5.3.3.

5.4.4 Effects on fresh concrete

(a) Unit mass
No effect.

(b) Workability
The retarder admixture keeps workability for a longer period.

(c) Slump loss
Most retarders are also water reducers. They increase initial and final setting times of the cement paste, and they increase the time required to reach the vibration limit. Therefore, it has been assumed that they also reduce slump loss. Laboratory and field experience demonstrate that water reducers/retarders do not substantially affect the absolute slump loss [26, 27, 31]. Distinction must be made between admixture addition at given water/cement ratio and at a given slump. Ramachandran [25] examined the effect of some water reducers/retarders on the slump loss of concrete at a given water/cement ratio. The initial slump of the concrete increased with the addition of the admixture. The rate of slump loss of the admixture-treated concrete was higher than that of the reference mix. Perenchio [17] found that, at a given initial slump, water reducers increased the rate of slump loss. The slump of the admixture-treated concrete was lower than that of the reference mix.

All these phenomena are affected by the cement composition and content, type, brand and dosage of admixture, ambient temperature, adding procedure of the admixture into the concrete, etc. It is possible to overcome the problem on the job site by planning ahead and design of the concrete mix based on all the data available concerning the specific job [3, 8, 9, 17, 20, 21].

(d) Bleeding

Retarders affect the rate and capacity of fresh concrete to bleed and settle under the influence of gravity. Different admixtures influence bleeding and settlement of concrete differently. For a given water/cement ratio [24] gluconate increases bleeding, glucose decreases it, and lignosulphonate has no effect on it. A higher reduction of bleeding is observed in the presence of air entraining agent [30]. The increase of partial mobility in presence of admixtures caused channelled bleeding and hence a great rate of settlement.

(e) Setting

Use of a retarding admixture in concrete generally causes a delay in initial and final setting times [23]. The specific retardation with a particular cement and different admixtures should be determined only by trial mixes [23]. All the commercial retarding admixtures should comply with the standards. A practical and simple test to determine the initial and final setting time of concrete is the Proctor penetration test, which is shown to be more efficient than the Pin-pull test [23].

By changing the dosage of the admixture, the vibration limit can be delayed for the desired length of time for an easy pouring of each layer of concrete to avoid cold joints. For any initial ambient temperature and concrete temperature, the dosage can be adjusted to maintain the desired time for the vibration of the concrete.

The retardation in initial and final setting times caused by a given amount of admixture, will probably be different at different temperatures. If the retarding effect is expressed in time, there will be a very small difference in the rate of setting between admixture-treated concrete and the reference mix.

(f) Heat of hydration and temperature

Heat of hydration and temperature rise of concretes containing retarders are less at early ages, and they equal at about 3–7 days. Sometimes they can be slightly higher than the reference mix at later ages. At a designed strength and slump, incorporation of water reducers decreases the heat of hydration and the temperature rise because of the reduction in the cement content [15].

(g) Plastic shrinkage
This generally occurs by loss of moisture in the first few hours (2–5 h) from placing and compacting of the concrete. If cracks are caused by plastic shrinkage, the retarding admixtures can aggravate the cracking problem.

5.4.5 Effects on hardened concrete

The effect of retarding admixtures on the properties of hardened concrete is very small. No further discussion on this issue will be done in this chapter.

5.4.6 Use

See Section 5.3.5.

5.4.7 Application

Retarders are often treated under the same category as "water reducers and retarders". This is because the main components used for retarders are also present in water reducing and retarding admixtures. As a result, many retarders reduce the water requirement and many water reducers retard the setting of the concrete.

Also, standard requirements for the initial and final setting times are approximately the same; though there are some differences in water reduction requirements and compressive strength.

Some of the more important engineering purposes for which retarders are used are:

* to retard time of initial and final setting;
* to reduce rate of slump loss;
* to retard or reduce heat evolution during early hardening;
* to increase vibration time.

5.4.8 Standards and code of practice

See Section 5.3.7.

5.5 REFERENCES

1. American Concrete Institute (1986), *ACI Manual of Concrete Practice*, Part 1.
2. Banfill, P. F. G., (1979) A discussion of the paper "Rheological properties of cement mixes", by M. Daimon and D. M. Roy, *Cement and Concrete Research*, Vol. 9, pp. 795–6.
3. Call, B. M., (1979) Slump loss with type "K" shrinkage compensating cement concretes and admixtures, *Concrete International*, Vol. 1, pp. 44–7.
4. Collepardi, M., Corradi, M., Baldini, G. and Pauri, M., (1980) Hydration of C_3A in the presence of lignosulphonate-carbonate system or sulphonated naphthalene polymer, *Proceedings of Seventh International Symposium on Chemistry of Cement*, Paris, Vol. IV, pp. 524–8.
5. Collepardi, M. and Pauri, M., (1982) Influence of lignosulphonate-carbonate system and naphthalene sulphonate on the hydration of C_3A and C_3A-Na_2O solid solution, *International Seminar on Calcium Aluminates*, Turin, Italy.
6. Cordon, A., (1960) Discussion of "Field experience using water-reducers in ready-mixed concrete", by E. L. Howard, K. K. Griffiths and W. E. Moulton, *ASTM Special Technical Publication*, No. 266, pp. 180–200.
7. Della Libera, G., (1967) Durability of concrete and its improvement through water-reducing admixtures, *International Symposium on Admixtures for Mortar and Concrete*, Brussels, Vol. 5, pp. 139–76.
8. Erlin, B. and Hime, W. G., (1979) Concrete slump loss and field examples of placement problems, *Concrete International*, Vol. 1, pp. 48–51.
9. Hersey, A. T., (1975) Slump loss, caused by admixture, *Journal of the American Concrete Institute, Proceedings*, Vol. 72, pp. 526–7.
10. Howard, E. L., Griffiths, K. K. and Moulton, W. E., (1960) Field experience using water-reducers in ready-mixed concrete, *ASTM Special Technical Publication*, No. 266, pp. 140–7.

11. Kalousek, G. L., (1974) Hydration processes at the early stages of cement hardening, Principal Paper, *Proceedings of the Sixth International Congress on the Chemistry of Cement*, Moscow.
12. Khalil, S. M. and Ward, M. A., (1978) Influence of SO_3 and C_3A on the early reaction rates of Portland cement in the presence of lignosulphonate, *Ceramic Bulletin*, Vol. 57, pp. 1116–22.
13. Khalil, S. M. and Ward, M. A., (1973) Influence of lignin based admixture on the hydration of Portland cements, *Cement and Concrete Research*, Vol. 3, pp. 677–88.
14. MacPherson, D. R. and Fisher, H. C., (1960) The effect of water-reducing admixtures and set retarding admixtures on the properties of hardened concrete, *ASTM Special Technical Publication*, No. 266, pp. 201–17.
15. Massazza, F. and Costa, U., (1980) Effect of superplasticizers on the C_3A hydration, *Proceedings of Seventh International Symposium on Chemistry of Cement*, Paris, Vol. IV, pp. 529–34.
16. Mielenz, R. C., (1968) Use of surface active agents, *Proceedings of the Fifth International Symposium on Chemistry of Cement*, Tokyo, Vol. 4, pp. 1–29.
17. Meyer, L. M. and Perenchio, W. F., (1979) Theory of concrete slump loss as related to the use of chemical admixtures, *Concrete International*, Vol. 1, pp. 36-43.
18. Odler, I., Schonfeld, R. and Dorr, H., (1978) On the combined effect of water solubles lignosulphonates and carbonates on Portland cement and clinker pastes, II. Mode of action and structure of the hydration products, *Cement and Concrete Research*, Vol. 8, pp. 525–38.
19. Pauri, M., Baldini, G. and Collepardi, M., (1982) Combined effect of lignosulphonate and carbonate on pure Portland clinker compounds hydration, II Tricalcium aluminate hydration, *Cement and Concrete Research*, pp. 271–7.
20. Polivka, M. and Klein, A., (1960) Effect of water-reducing and retarding admixtures as influenced by Portland cement composition, *ASTM Special Technical Publication*, No. 266, pp. 124–39.
21. Previte, R. W., (1977) Concrete slump loss, *Journal of the American Concrete Institute, Proceedings*, Vol. 74, pp. 361–7.
22. Prior, M. E. and Adams, A. B., (1960) Introduction to Producers' papers on water-reducing admixtures and set-retarding admixtures for concrete, *ASTM Special Technical Publication*, No. 266, pp.

170–9.
23. Ramachandran, V. S., (1984) *Concrete Admixture Handbook: Properties, Science and Technology*, Noyes Publications, USA, Chapter 3, pp. 116–201.
24. Ramachandran, V. S., (1978) Effect of sugar-free lignosulphonates on cement hydration, *Zement –Kalk –Gips*, Vol. 31, pp. 206–10.
25. Ramachandran, V. S., (1981) Effect of retarders/water reducers on slump loss in superplasticized concrete, *Development in the Use of Superplasticizers*, American Concrete Institute, ACI SP-68, pp. 393–407.
26. Rossington, D. R. and Runk, E. J., (1968) Adsorption of admixtures on Portland cement hydration products, *Journal of American Ceramic Society*, Vol. 51, pp. 46-50.
27. Tuthill, L. H., (1979) Slump loss, *Concrete International*, Vol. 1, pp. 30–55.
28. Tuthill, L. H., Adams, R. F. and Hemme, J. H. Jr., (1960) Observations in testing and use of water-reducing retarders, *ASTM Special Technical Publication*, No. 266, pp. 97-117.
29. Tuthill, L. H. and Cordon, W. A., (1955) Properties and uses of initially retarded concrete, *Journal of American Concrete Institute, Proceedings*, Vol. 52, No. 3, pp. 273–86.
30. Vollick, C. A., (1960) Effect of water-reducing admixtures and set-retarding admixtures on the properties of plastic concrete, *ASTM Special Technical Publication*, No. 266, pp. 180–200.
31. Wallace, G. B. and Ore, E. L., (1960) Structural and lean mass concrete as affected by water-reducing, set-retarding agents, *ASTM Special Technical Publication*, No. 266, pp. 38–94.
32. Young, J. F, (1972) A review of the mechanisms of set-retardation in Portland cement pastes containing organic admixtures, *Cement and Concrete Research*, Vol. 2, pp. 415–33.

PART TWO

OTHER ADMIXTURES

6

Antiwashout admixtures for underwater concrete

S. Nagataki

6.1 DEFINITION

Antiwashout admixtures for underwater concreting reduce segregation of the concrete due to increasing cohesiveness.

6.2 INTRODUCTION

When concrete is being placed underwater, the cement may be washed out from the concrete, with consequent reduction of the strength of the concrete and pollution of the surrounding water. In order to resolve these problems, antiwashout admixtures composed mainly of water-soluble polymer are used. Concrete containing antiwashout admixture is a new type of underwater concrete which does not easily segregate even when placed directly under water [1].

6.3 COMPOSITION

Antiwashout admixtures for underwater concrete can be divided into two types [2]. One has water-soluble cellulose ether [2, 3] as its main component, the other has water-soluble acryl-type polymer [4] as its main component.

Chemical formula of water-soluble cellulose ether

Chemical formula of water-soluble acryl-type polymer

6.4 MECHANISM OF ACTION

These admixtures increase the cohesiveness of the concrete by increasing the viscosity of water [3] in the concrete and thereby improving the resistance to segregation. The resistance to segregation depends mainly on the dosage and molecular weight of the admixture.

6.5 EFFECTS ON CONCRETE IN THE FRESH STAGE

6.5.1 Workability

Concrete containing antiwashout admixture has high cohesiveness compared with ordinary concrete. The concrete containing the admixture has excellent self-levelling properties [2, 3, 5]. This property appears in the slump test. Slump increases gradually for about 10 minutes and the final slump value for a range of practical values is 23–26 cm as shown in Fig. 1 [5].

Immediately after pulling up slump cone

Three minutes after pulling up slump cone

Fig. 1. Slump test

6.5.2 Segregation

When a concrete containing antiwashout admixture is poured under water by free fall, it is hardly segregated, as shown in Fig. 2 [5]. But ordinary concrete without the admixture is completely segregated.

The results of falling through water test to evaluate resistance to segregation are shown in Figs 3 and 4. pH value and turbidity of the water decreased with the increase in dosage of the admixture [2, 4, 5].

Underwater concrete containing high fineness slag has excellent resistance and strength developing properties [9].

The method of the falling-through-water test is as follows: 3 litres of concrete are allowed to fall through water in a container of 30 cm diameter and 50 cm height. The Ph value and turbidity of the water near the surface are measured 60 seconds later.

Ordinary concrete   Concrete containing
                                     antiwashout admixture

Fig. 2. Underwater segregation resistance test

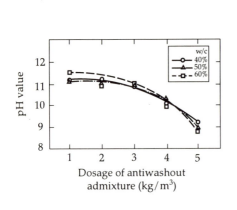

Fig. 3. pH value

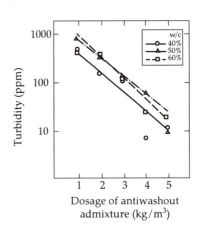

Fig. 4. Turbidity

6.6 EFFECTS ON CONCRETE IN THE SETTING STAGE

6.6.1 Setting

Concrete containing cellulose ether tends to delay setting time [2, 3, 5, 6]. Concrete containing acryl-type polymer hardly affects setting time [8], but when using with water reducer and/or superplasticizer, it tends to have a delayed setting time.

6.6.2 Bleeding

Concrete containing an antiwashout admixture has a high water retentivity due to viscosity increase and causes no bleeding at all even when a small dosage of the admixture is added [2].

6.7 Effects on concrete in the hardened stage

Strength [2–10]: As shown in Fig. 5, the strength ratio f^1w/f^1a (f^1w and f^1a are compressive strength at the age of 28 days of specimens cast in water and in air respectively) increased with the increase in dosage of antiwashout admixture due to high resistance to segregation in water [5].

6.8 USE

Most antiwashout admixtures are dosed in the range 1–1.5% by weight of water (unit weight 2–3 kg/m^3) and are frequently used in combination with superplasticizers [3].

Antiwashout admixture are usually discharged into the mixer at the same time as other materials (cement, aggregate and water).

6.9 APPLICATIONS

Concrete containing antiwashout admixture for underwater concrete is very effective for the following uses:

- foundations of bridge piers in water and slabs;
- revetments, piled piers and breakwaters of ports;
- prevention of scouring near underwater structures and protection of river beds;
- concrete for reinforcing rubblestone of breakwaters;
- reinforced concrete for slabs of caissons.

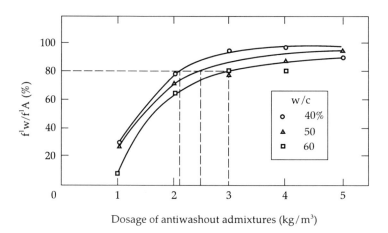

Fig. 5   Relationship between dosage of antiwashout admixture and f^1w/f^1A

6.10 SPECIFICATIONS

Specifications of concrete containing antiwashout admixture for underwater concrete in Japan are shown in Table 1.

Table 1. Specifications of concrete containing antiwashout admixture for underwater concrete [11].

Item		Standard type	Retarded type
Bleeding (%)		< 0.01	< 0.01
Air (%)		< 4.5	< 4.5
Slump flow loss (cm)		< 3.0 After (30 min)	< 3.0 After (2 h)
Initial setting time (h)		> 5	> 18
Final setting time (h)		< 24	< 48
Compressive strength of specimen cast in water (kg/cm^3)	7 days	> 150	> 150
	28 days	> 250	> 250
Strength of specimens cast in water Strength of specimens cast in air (%)	7 days	> 80	> 80
	28 days	> 80	> 80
Falling through water test	SS (mg/l)	< 50	< 50
	pH	< 12.0	< 12.0

6.11 REFERENCES

1. Japan Society of Civil Engineers, (1990) *Proceedings of Symposium on Antiwashout Underwater Concrete*, Tokyo, August. (The full contents list of this publication appears as an appendix to this chapter.)
2. Sogo, S., Haga, T. and Nakagawa, T., (1987) Underwater concrete containing segregation controlling polymers, *Fifth International Congress on Polymers in Concrete*, Brighton, England, September.
3. Kawai, T., (1987) Non-dispersible underwater concrete using polymers, *Fifth International Congress on Polymers in Concrete*, Brighton, England, September.
4. Sakuta, M., Yoshioka, Y. and Kaya, T., (1983) Use of acryl-type

polymer as admixture for underwater concrete, *Polymers in Concrete*, American Concrete Institute.

5. Nakajo, N., (1987) Non-dispersible underwater concrete mixed with cellulose ether based polymer, *Fifth International Congress on Polymers in Concrete*, Brighton, England.
6. Tamada, S., Sogo, S. and Miura, N., (1985) Properties of underwater jointed concrete containing segregation controlling admixture, *CAJ Review*.
7. Tazawa, Y., Ohtomo, T. and Taira, K., (1989) Properties of antiwashout concrete with high blast furnace slag content, *Third International Conference on The Use of Fly Ash, Silica Fume, Slag and Natural Pozzolanas in Concrete*, American Concrete Institute, ACI SP-114.
8. Tachihata, S. et al, (1983) Strength of underwater concrete using water retaining admixture, *CAJ Review*.
9. Kagaya, M., Tokuda, H. and Tsutaho, K., (1986) Segregation characteristics of underwater concrete, *CAJ Review*.
10. Hara, M. et al, (1990) Admixing effect of high fineness slag on the properties of underwater concrete, *Admixtures for Concrete, Improvement of Properties*, E. Vazquez, editor, Chapman & Hall, London, pp. 440-8.
11. Coastal Development Institute of Technology, Japanese Institute of Technology on Fishing Ports and Communities, (1986) *Manual for Special Underwater Concrete*. (In Japanese.)
12. Japan Society of Civil Engineers, (1991) *A Standard for the Quality of Antiwashout Admixture for Underwater Concrete*.

Appendix: Proceedings of the Symposium on Antiwashout Underwater Concrete, Tokyo, August 1990, Japan Society of Civil Engineers.

1. Chung, Y., Tsurumi, T., Asaga, K. and Daimon, M., Rheological property of cement paste with a segregation controlling admixture, pp. 1–6.
2. Shioya, M., Saitou, T., Kikutani, T., Otuki, N. and Tataku, A., Viscous-flow behaviour and microstructure of cement paste with antiwashout admixtures, pp. 7–14.
3. Hayakawa, K., Yamakawa T. and Otsuki, N., Fundamental studies on the distribution and the dissolution of antiwashout admixtures, pp. 15-20.
4. Yamakawa, H., Yoshioka, Y., Itoh, K. and Shioya, T., Concrete properties with acryl-type polymers as antiwashout, pp. 21–8.
5. Kakuta, S., Asano, F., and Kojima, T., Study on the consistency of antiwashout underwater concrete, pp. 29–36.
6. Kokuba, K., Effects of cements and admixtures on fluidity, setting and strength development of antiwashout underwater concrete, pp. 37–44.
7. Tazawa, E., Yonekura, A., Kasai, T. and Tugawa, K., Effect of aggregates and admixtures on fluidity and strength of antiwashout concrete, pp. 45–52.
8. Nakagawa, Y., Ohtomo, T., Nakahira, J. and Matsuoka, Y., Study on antiwashoutability, setting and long-term flowability of antiwashout underwater concrete, pp. 53–60.
9. Tatekawa, T., Halada, A., Sogo, S., Takeda, N. and Yamakawa, T., An experiment for producing method of antiwashout underwater concrete in site, pp. 61-8.
10. Matumuro, Y., Yasuda, M., Isizuka, K., Takimoto, H. and Yasui, S., Development of site remixing system for antiwashout underwater concrete, pp. 69-76.
11. Nakahira, J., Nakagawa, Y., Ohtomo, T., and Matsuoka, Y., Study on mixing method of antiwashout underwater concrete, pp. 77–84.
12. Kawakami, M., Abe, M., Kimura, M. and Takemoto, T., On the propriety of the method of making colloidal underwater concrete specimens, pp. 85–92.
13. Ohkura, M., Kobayashi, S., Morihama, K., and Takahashi, H., Experiment of strength and durability for antiwashout underwater

86 *Application of admixtures in concrete*

concrete, pp. 93–100.

14. Matsushita, H., Yamato, T. and Emoto, Y., Compressive fatigue properties of antiwashout underwater concrete, pp. 101–6.

15. Takagi, N., Ono, K. and Miyagawa, T., Creep of antiwashout underwater concrete under water, pp. 107–14.

16. Nagataki, S. and Inoue, T., Thermal properties of antiwashout underwater concrete, pp. 115–22.

17. Ohtomo, T., Nakagawa, Y., Nakahira, J. and Matsuoka, Y., Study on effect of antiwashout admixture on hardened antiwashout underwater concrete, pp. 123–30.

18. Seki, H., Miyata, K., Kitamine, H. and Kaneko, Y., Experimental study on watertightness of underwater mortar using antiwashout agents, pp. 131–6.

19. Influence of antiwashout admixtures on chloride penetration in mortar

20. Fukute, T. and Hamada, H., A study on the corrosion of steel bars embedded in the antiwashout underwater concrete – longterm (6 years) exposure test under marine environment, pp. 145–52.

21. Motohashi, K., Mizobuchi, T. and Suda, K., Effect of marine exposure on behaviour of RC beams cast with antiwashout concrete, pp. 153–60.

22. Yamato, T., Emoto, Y. and Soeda, M., Freezing-and-thawing resistance of antiwashout concrete under water, pp. 161–6.

23. Chouai, T. and Yamamoto, Y., A study on the freeze-thaw durability of antiwashout underwater concrete, pp. 167–74.

24. Fukudome, K., Kita, T., Miyano, K. and Taniguchi, H., Property varies of antiwashout underwater concrete by flowing under water, pp. 175–80.

25. Ohtomo, T., Motohashi, K., Izumi, T. and Tazaki, K., Effect of underwater flow on properties of antiwashout concrete, pp. 181–8.

26. Sano, K. and Sueoka, E., On flowability and quality of antiwashout underwater concrete, pp. 189–94.

27. Taniguchi, H., Fukudome, K., Miyano, K. and Kita, T., Effect of flowing water on the behaviour of antiwashout underwater concrete, pp. 195–202.

28. Watanabe, T., Shuttoh, K., Ohtake, H. and Satoh, F., The study on placing of antiwashout underwater concrete in flowing water, pp. 203–8.

29. Wakamatsu, G., Minami, T., Ozawa, I., Chikamatsu, R. and Sogo,

S., Performance of antiwashout underwater concrete placed in running water, pp. 209-16.

30. Kasima, S., Higuchi, K. and Kitaguchi, M., Fluidity and quality after flowing of antiwashout underwater concrete, pp. 217–22.
31. Takahasi, N. and Hosaka, T., Construction of bridge foundation 20 metres below sea level using antiwashout underwater concrete - access bridge to the Kansai International Airport, pp. 223–8.
32. Iwaki, I., Ando, K. and Shindo, T., Antiwashout underwater mortar (shear key) used in terminal block of submerged tunnel, pp. 229–34.
33. Sakurai, H., Ando, K. and Yoshihara, M., The development of lightweight back filling material on underwater construction, pp. 235 ff.

7

Corrosion inhibitors

E. Vazquez

7.1 DEFINITION

A corrosion inhibitor is a chemical compound added to reinforced concrete, to delay corrosion of the steel.

7.2 INTRODUCTION

The corrosion resistance of reinforcement is affected adversely by chloride ions and by the decrease in the pH of the concrete pore solution. The main factor influencing this decrease is the carbonation of the concrete cover by atmospheric carbon dioxide. Low permeability of concrete is the best protection for the reinforcement. But in some cases when supplementary protection of the reinforcement is needed, it is advisable to treat the steel or the concrete. The addition of corrosion inhibitors to concrete can increase the protection of the steel.

7.3 COMPOSITION

Corrosion inhibitors can be divided in three classes

1. anodic
2. cathodic
3. mixed.

These interfere with corrosion reactions at anodic or cathodic sites, or both.

For anodic inhibitors the most commonly used materials are calcium and sodium nitrite. In some countries calcium nitrite is the only corrosion inhibitor available. Sodium benzoate and sodium chromate can also be used.

Most cathodic inhibitors are investigated only on a laboratory scale. The principal work has centred on aniline and its chloro alkyl-and nitro- substituted forms.

Mixed inhibitors contain one or two types of molecules with proton and electron acceptor groups. Both groups can be in the same molecule (aminobenzenethiol) or salt can be formed by the different orienting groups of two separate molecules.

7.4 MECHANISM OF ACTION

Chlorides may be present in concrete derived from the aggregates, cement or water or they can penetrate from the outside.

In the presence of chloride the protective passivating layer of the steel is breached. The passivating layer formed in the surface is destroyed by these reactions:

$$\text{Fe} \rightarrow \text{Fe}^{++} + 2e^- \text{ (anode)} \qquad \overset{2e}{\frac{1}{2}} \text{O} + \text{H}_2\text{O} \rightarrow 2\text{OH}^- \text{ (cathode)}$$

$$\text{Fe}^{++} + 2\text{OH}^- \rightarrow \text{FeO}.n(\text{H}_2\text{O})$$

This takes place because Fe^{++} in the form of chloride complex migrate away, exposing the steel to corrosion. The chloride complex is further oxidized to different Fe^{+++} compounds:

$$\text{Fe}^{++} + \text{Cl}^- + \text{OH}^- \rightarrow (\text{Fe Cl})^+ \ \text{OH}^-$$
$$\text{soluble chloride complex}$$
$$\downarrow \text{H}_2\text{O}$$
$$\text{O}_2$$
$$(\text{Fe(OH)}_3, \text{Fe}_2\text{O}_3, \text{Fe}_3\text{O}_4).n(\text{H}_2\text{O})$$

The corrosion products have a larger volume than that of the original steel, and pressures build up that exceed the tensile strength of the concrete, causing cracking.

With the addition of nitrite (Na^+ or Ca^+) to the concrete a competing oxidation reaction is initiated at the surface of the steel which regenerates the passivating layer with Fe_2O_3:

$$2Fe^{++} + 2OH^- + \underset{\substack{\text{anodic} \\ \text{inhibitor}}}{2NO_2^-} \rightarrow 2NO + Fe_2O_3 + H_2O$$

The carbonation of concrete by atmospheric CO_2

$$CO_2 + Ca^{++} + 2OH^- \rightarrow CaCO_3 + H_2O$$

dramatically decreases the pH values, and corrosion potentially reaches active values. When NO_2^- is added the change in pH values is the same, but the corrosion potential remains in passive values. So, the NO_2^- avoids corrosion as much in alkaline as in neutral media. Nitrite ions rapidly oxidize ferrous ion to ferric, blocking further passage of ferrous ion from steel into the electrolyte.

Some authors have reported that NO_2^- oxidizes in concrete to NO_3^- reducing the time of action of the corrosion inhibitor. Other authors have reported that NO_2^- acts on steel only as an oxidant and it reduces.

Nitrites, like other substances such as chlorides, combine with C_3A and C_4AF, and their effective amount in concrete is less than that added, because the only effective amount is that which can reach the steel surface. Their effectiveness also depends on their proportion in relation to the amount of chloride and the ambient humidity.

More calcium nitrite will protect against more chloride. If the ratio of inhibitor : chloride ratio is small the protective film repair by the nitrite ion and chloride attack occur simultaneously, corrosion will dominate and become localized.

Cements containing higher levels of C_3A offer better corrosion resistance.

(a) Cathodic inhibitors
These act by slowing the cathodic reaction or selectively precipitating on cathodic sites to increase the circuit resistance and restrict the diffusion of reducible species to the cathode.

(b) Mixed inhibitors
These contain molecules in which electron density distribution causes the inhibitor to be attracted to both anodic and cathodic sites.

7.5 EFFECTS ON CONCRETE

7.5.1 Fresh state

Workability Workability is slightly increased by the inorganic salts used as corrosion inhibitors. Organic inhibitors seem not to affect workability.

7.5.2 Setting stage

Setting Inorganic admixtures, as nitrites, accelerate initial and final set times. With some corrosion inhibitors, if the acceleration is not desired, a retarder can be added.

In general terms, the effect of this admixture on properties of concrete in the setting stage are the same as accelerators.

7.5.3 Hardening stage

(a) Strength
Calcium nitrite slightly increases compressive strength. Sodium nitrite has no effect on compressive strength at early ages and decreases 28 day strength slightly. The same tendency is observed for chromates.

Sodium benzoate significantly reduces concrete strength at all ages. Similar effects are observed for flexural strength.

(b) Bond
In some cases the inhibiting layer formed by the inhibitor on the steel surface decreases the bond strength between steel and concrete.

7.5.4 Hardened state

(a) Alkali–aggregate reaction
Inhibitors based on sodium salts such as sodium nitrite and sodium benzoate may increase the effects of alkali–aggregate expansion if reactive aggregates are present.

(b) Efflorescence
Inhibitors based on sodium salts such as sodium nitrite and sodium benzoate can increase efflorescence.

7.6 USE

7.6.1 Calcium nitrite

Calcium nitrite is the most used corrosion inhibitor.   It is also an accelerator.   The admixture solution contains about 30% calcium nitrite solid by weight.   It is normally added to concrete during the batching process and is dispersed throughout the concrete.   As a general rule, a minimum amount of 3% by weight of cement is recommended.   In Japan it is used at rates from 0.9 to 1.2 kg (solid weight) per unit volume (m^3) of concrete.

The protection given by the inhibitor depends on the quality of the concrete.   The length of time of protection may be short in poor quality concrete.   Calcium nitrite improves the resistance of concrete to chloride attack at water/cement ratios under 0.5.   Some authors report that when water reducing agents were used, the time of corrosion protection could be extended.   So calcium nitrite is frequently used in combination with superplasticizers.

In air entrained concrete the addition of calcium nitrite requires higher dosages of air entraining agent to keep the air content at a given level.   Curing at elevated temperature or alternate wetting and drying promotes corrosion and reduces the effectiveness of calcium nitrite.

Most authors report that calcium nitrite does not completely avoid corrosion but it always reduces it significantly allowing a lengthening of the service life.

7.6.2 Sodium nitrite

In the absence of chlorides sodium nitrite is used effectively at dosages of 1-2% by weight of cement. In the presence of chlorides the dosage should be higher. Sodium nitrite is less used because of the disadvantages associated with the presence of Na^+.

7.6.3 Sodium and potassium chromates

They are used at 2-4% by solid weight of cement. The effects are similar to those of the nitrites. Potassium chromate produces a light green colour of concrete.

7.6.4 Sodium benzoate
It is seldom used but some authors found its effect to be more persistent than other admixtures mentioned.

7.6.5 Stannous chloride and hydrazine hydrate
These have only been investigated and found by some authors to be efficient.

7.6.6 Cathodic and mixed inhibitors
These are less used. Dosage is generally at the level 1-2% by cement weight.

7.7 APPLICATIONS

Calcium nitrite has been used in reinforced concrete and prestressed concrete structures exposed to chloride attack.

Sodium benzoate has been used in some cases in the UK, and sodium nitrite in different applications in Europe.

7.8 SPECIFICATIONS

The only nationally recognized standard is the Japan Industrial Standard A-6205 'Corrosion inhibitor for reinforcing steel in concrete'.

As electrochemical measurements including potential and polarization measurements of steel in concretes are the best methods of determining corrosion, they can be used in the evaluation of corrosion inhibitors. The use of open-circuit potential measurements is considered in the Special Technical Publication 713 of ASTM (1980). The main reference for test methods are:

Rapid Method of Studying Corrosion Inhibition of Steel in Concrete. Portland Cement Association Research and Development Bulletin 187, PCA, Skokie, Illinois.

Among other interesting publications the following may be mentioned:

Berke N.S (1986) The use of anodic polarization to determine the effectiveness of calcium nitrite as an anodic inhibitor, ASTM STP 906, Philadelphia.

8

Gas forming admixtures

D. Dimic

8.1 DEFINITION

Gas forming admixtures generate or liberate bubbles of gas in the fresh mixture during the hydration process before the initial set of the cement paste matrix takes place [1, 2, 3].

8.2 INTRODUCTION

Gas forming admixtures are used for a number of different purposes:

- to control settlement and bleeding [4–7],
- to improve the intrusion of grouts and mortars in preplaced-aggregate concrete, by producing an expansion of 5–10% [8], or even up to 14% [9],
- for the production of aerated concrete and mortars [10].

8.3 COMPOSITION

The components used as gas forming admixtures are metallic aluminium, zinc or magnesium, hydrogen peroxide, nitrogen and ammonium compounds, and certain forms of activated carbon or fluidized coke [1, 2, 10]. The nitrogen compounds used are azo or hidrazin compounds with at least one N–N bond in the molecule.

8.4 MECHANISM OF ACTION

When a gas forming admixture is added to a grout, mortar or concrete, a reaction between the gas forming admixture and the components present in the fresh hydraulic cement mix takes place. When a known quantity of the gas forming admixture is added, meaning the liberation of a known volume of gas, a predetermined rate of expansion can be expected. For a given application the rate of expansion and its duration are important factors for the properties of the mortar or concrete. The homogeneity of the distribution and size of the bubbles generated by the reaction are important factors, too.

Hydrogen, oxygen, nitrogen or air are released by the chemical reactions. Hydrogen gas is produced by the reaction of calcium hydroxide, liberated by cement hydration, with aluminium powder. The gas formation which occurs in this case may be illustrated by the following simplified equation [10]:

$$2Al + 3Ca(OH)_2 + 6H_2O \rightarrow 3CaO.Al_2O_3.6H_2O + 3H_2$$

If hydrogen peroxide and calcium hypochloride (bleaching powder) are used, a reaction in which oxygen is evolved instead of hydrogen takes place.

The calcium chloride released by this reaction acts as an accelerator for the cement hydration process; however, the reader is cautioned to evaluate the corrosion potential of the calcium hypochloride and calcium chloride.

Nitrogen gas is produced by the decomposition of the N–N bond of the compound, influenced by the action of activators. Activators like aluminates or copper salts are used.

Certain forms of activated carbon or fluidized coke, with a moisture content of about 3%, liberate adsorbed air [2].

8.5 MAIN FACTORS INFLUENCING GAS EVOLUTION

The following factors affect the rate and duration of the gas evolution.

8.5.1 Mix proportions

The amount of water and the consistency of the fresh mix influence the rate and continuity of the hydrogen liberation. There is an optimal amount of water and an optimal consistency of the grout, mortar or concrete at which maximum expansion is obtained. At highly plastic or flowable consistencies a significant amount of gas may escape from the mix before setting [11, 12, 13]. Use of a thickener will increase expansion efficiency by preventing the gas from escaping [14].

8.5.2 Characteristics of the cement

The fineness and alkali content of the cement are important for the gas evolution. Gas evolution accelerates as cement fineness and alkali content increase. The activity and relative content of the minerals, particularly tricalcium silicate, influence the speed of hydrogen release and the duration of the expansion [2, 9]. The type of cement also affects gas generation rates and the expansion starting time [15].

The reactions when nitrogen or oxygen are liberated are independent of the cement characteristics, as they are initiated and controlled by activators.

8.5.3 Characteristics of the admixture

Aluminium powder may be used in unpolished or polished form, depending on the desired speed of the gas evolution. Unpolished aluminium powder is generally preferred. It reacts faster, whereas polished powder is used when a slower reaction is desired. Aluminium powder with a specially treated surface is also used for reaction control [15, 16].

The fineness and particle shape of the aluminium powder also influence the rate and the duration of gas evolution and have an influence on the size and shape of the air pores in the hardened mix [1, 10].

When using fluidized coke, the fineness and moisture content are important factors for the gas evolution process [2].

8.5.4 Temperature

The ambient temperature and the temperature of the mix influence hydrogen generation considerably. Nitrogen, oxygen or air producing systems are not directly influenced by temperature [2].

The reaction of aluminium powder with calcium hydroxide is faster at higher temperatures, so higher dosages of admixture are used at lower temperatures to ensure the same rate of expansion as is reached at normal ambient or mix temperatures.

It is possible to speed up the rate of gas generation in cold weather by the addition of accelerators such as sodium hydroxide, hydrated lime or trisodium phosphate [2, 10]. This may ensure sufficient gas generation before setting.

8.6 EFFECTS

8.6.1 Fresh stage

(a) Unit weight
The unit weight of the grout, mortar or concrete is reduced, depending on the degree of expansion, the degree of restraint, and the stability of the bubbles.

(b) Workability
Gas forming admixtures have no significant influence on workability or workability loss.

(c) Volume changes
When properly predetermined, the admixture produces expansion which compensates for the volume change caused by settlement and plastic shrinkage, and causes a slight expansion of the freshly mixed grout, mortar or concrete.

8.6.2 Setting stage

Gas forming admixtures do not alter the initial and final setting times of the mixes. Retarders are often used for grouting long ducts or rock anchors and for oil well grouting, where high temperatures and pressures may be encountered and the distances are relatively great [1, 2, 15, 16].

8.6.3 Hardened stage

(a) Strength
Strengths are reduced to some extent [11]. The reduction in strength depends on the degree to which the expansion of the mixture is restrained; little reduction in strength may be achieved if suitably restrained. If the placed mix is not fully restrained, the gas bubbles tend to be displaced upwards and are therefore not uniformly dispersed throughout the hardened mix. In this case the reduction of strength may be very significant. It is therefore important that the forms be tight and well sealed.

Water reducers or high-range water reducers present in the multicomponent admixtures may be able to compensate or modify loss of strength.

(b) Volume changes
Gas forming admixtures do not overcome the shrinkage caused by drying or carbonation. Usually the drying shrinkage of grouts, mortars or concretes containing gas forming admixtures is increased [12].

(c) Durability
Gas forming admixtures are not used specifically to improve durability [12, 17, 18].

8.7 APPLICATION

Aluminium powder as extensively used is normally added at rates between 0.006% and 0.02% by weight of the cement, although larger quantities may be used to produce lightweight concrete. Higher amounts are required when the temperature is lower. Commercially available products are generally premixed with fine powder materials (sand, cement, pozzolana) because of the very small quantities of the aluminium powder generally used.

As gas forming admixtures are commonly used for specific purposes like structural grouting and the cementing of oil wells, they are usually produced as a combination of a gas forming agent and a plasticizing agent, retarder, accelerator or other materials in order to impart the special properties desired.

Preblended mixtures of cement, sand or filler, gas forming and other admixtures are available on the market as ready-to-use grouts with specific properties.

Gas forming admixtures as single or multicomponent admixtures are used to produce the following products.

(a) Grouts, mortars or concretes with controlled expansion
The controlled expansion avoids excess pressure build up, but sufficient expansion is developed to provide good bond of the fresh mix to the contact surface, with the reduction of voids [4, 6, 8, 9, 19, 20, 21]. They can be used for:

- the grouting of cable ducts in post-tensioned concrete,
- the setting up of heavy machinery,
- the fixing, jointing and anchoring of precast concrete elements,
- repair work,
- the grouting of joints between inversely placed concrete elements.

(b) Intrusion grouts or mortars for preplaced aggregate concrete
The admixtures used for this special type of grout or mortar, also called intrusion aids, shall comply different requirements; up to 14% in US [9]; in Japan the expansion should be in the range from 5 to 10% [8]. This concrete is produced preferably as mass concrete, for underwater concreting and for placing heavyweight concretes.

(c) Cellular concrete
Of the various methods in which air cells may be formed in the concrete, the one of chief practical importance today is the aluminium powder process [10].

8.8 SPECIFICATIONS

At present, the use, sampling and testing of gas forming admixtures is covered by various standards, guides or recommendations. The table overleaf gives some of them.

8.9 REFERENCES

1. American Concrete Institute Committee 212, 3R, (1989) *Admixtures for Concrete.*
2. Ramachandran, V. S., (1984) *Concrete Admixtures Handbook: Properties, Science and Technology*, Noyes Publications, New Jersey, USA.
3. Russell, P., (1983) *Concrete Admixtures*, Viewpoint Publications (E & FN Spon), London.
4. Prestressed Concrete Institute Committee, (1972) Recommended practice for grouting of post tensioned concrete, *PCI Journal*, November/December.
5. Report on Grout and Grouting of Prestressed Concrete, (1974) *Proceedings of the Seventh Congress of the Federation Internationale de la Precontrainte* .
6. American Concrete Institute Committee 301, (1972) *Specification for Structural Concrete for Buildings.*
7. Schupack, M., (1974) Development of a water retentive grouting aid to control the bleed in cement grout used for post-tensioning, *Proceedings of the Seventh Congress of the Federation Internationale de la Precontrainte.*
8. Japan Society of Civil Engineers, *Standard Specifications for Design and Construction of Concrete Structures (Construction).*
9. American Society for Testing and Materials, (1989) Standard specifications for grout fluidifer for preplaced-aggregate concrete, ASTM C 937-80, *Annual Book of ASTM Standards*, Vol. 04.02, pp. 460-1.
10. Short, A. and Kinniburgh, W., (1963) *Lightweight Concrete*, John Wiley and Sons, NY, USA.
11. Menzel, A. A., (1943) Some factors influencing the strength of concrete containing admixtures of powdered aluminium, *Journal of the American Concrete Institute*, Vol. 39, No. 3, pp. 165-84.
12. Paillere, A. M., (1971) Etude comparative de quelques coulis d'injection, *Bulletin Liaison Laboratoire des Ponts et Chaussées*, No. 52, Ref. 1059.

Australia	AS-2073-79	Methods for testing of expanding admixtures for concrete, mortar and grout
	AS-2072-79	Methods for sampling of expanding admixtures for concrete, mortar and grout
USA	CRD C 621-83	Corps of Engineers Specification for nonshrink grouts
	CRD C 619	Corps of Engineers Specification for grout fluidifier
	ASTM C 937-80	Standard specification for grout fluidifier for preplaced-aggregate concrete
	ASTM C 938-80	Standard specification for proportioning grout mixtures for preplaced-aggregate concrete
	ASTM C 1090-88	Standard test method for measuring, changes in height of cylindrical specimen from hydraulic cement grout
	ASTM C 1107-89	Standard specification for packaged dry, hydraulic-cement grout (nonshrinkable)
Great Britain	BS 5400: 1978	Steel, concrete and composite bridges. Grouting of prestressing tendons. Clause 3.12, Part 7 and Part 8
Switzerland	SIA-Norm 162	Norm für die Berechnung, Konstruktion und Ausfuhrung von Bauwerken am Beton, Stahlbeton und Spannbeton
Austria	ÖNORM B 4250	Spannbetontragwerke, Berechnung und Ausfuhrung, 1974
France		Directive provisoire sur les injections des graines des ouvrages en beton precontraint
Germany	DIN 4227 Teil 5	Spannbeton. Einpressen von Zementmartel in Spannkanäle
Yugoslavia	JUS U.E3.015.86	Grouts for use in bonded posttensioned prestressed concrete

13. Volter, O., (1962) Die veranderlichkeit des einpressmortels auf der baustelle, *Beton und Stahlbetonbau*, Nr. 10, pp. 239-44.
14. Nagataki, S., Tokyo Institute of Technology, Private Communication.
15. Taisei Technical Report, (1983) *CONEC Method*.
16. Takase et al, (1985) Inverted casting method of concrete wall using Al-flake, *Proceedings JSCE*, No. 335/VI-1.
17. MacInnes, C., (1968) The frost resistance of cement grouts for prestressed concrete applications, *Proceedings of the Fifth International Symposium on the Chemistry of Cement*, Tokyo, Part III, pp. 260-73.
18. Ronisch, A., (1955) Die einwirkung von frost auf den einpress-mortel von spanngliedern, *Beton und Stahlbetonbau*, No. 2, pp. 64-71.
19. Vorlaufiges Merkblatt fur Zementeinpressungen im Bergbau, Fassung, (1970) *Beton*, 20, No. 1, pp. 19-22.
20. Rendchen, K., (1977) Einpressmörtel für spannbeton-vergleich der vorschriften und empfehlung verschiedener länder, *Betontechnische Berichte, Beton*, No. 11, 12, pp. 437-43, 447-82.
21. American Concrete Institute Committee 304, (1991) Guide for the use of preplaced aggregate concrete for structural and mass concrete applications, *ACI Materials Journal*, Vol. 88, No. 6, pp. 650-68.

9

Pumping aids

E. Vazquez

9.1 DEFINITION

Pumping aids are admixtures that enhance the pumpability of concrete.

9.2 INTRODUCTION

Fresh concrete for pumping must be sufficiently mobile, cohesive enough not to segregate and capable of flowing with reduced friction along pipe walls. Air entraining admixtures and water reducers improve flow at normal cement dosages. Superplasticizers are more effective in richer concretes. These admixtures are excluded in this chapter from the category of pumping aids. Pumping aids considered in this state-of-the-art report are admixtures used to pump marginally pumpable concrete. They improve the plastic properties of concrete in special situations.

9.3 COMPOSITION

Pumping aids usually consist of the following materials:

*   synthetic and natural organic polymers, soluble in water. In this class are cellulose ethers, polyethylene oxides, carboxyl vinyl polymers, polyvinyl alcohol and others,
*   dispersions of natural gums, styrene copolymers with carboxyl groups and synthetic polyelectrolytes,

- acrylic emulsions.

According to the definition of admixtures, products such as condensed silica fume, fly ash, kaolin, rock dust and pozzolanic materials are not included in this report.

9.4 MECHANISM OF ACTION

These products act by physical means on concrete properties.

The admixture should lubricate the fresh concrete to give an increase in cohesiveness, and reduce bleeding, especially in low cement content mixes.

So-called thickeners increase the viscosity of the mixing water.

Dispersions of natural gums, styrene copolymers with carboxyl groups and synthetic polyelectrolytes are products that increase viscosity by promoting interparticle attraction. They are flocculants which are adsorbed on the cement particles.

Acrylic emulsions are products that increase viscosity by interparticle attraction and also by supplying superfine particles in the cement paste.

9.5 EFFECTS

9.5.1 Fresh state

Workability The effect of a pumping aid admixture on the workability of concrete depends upon the cement type and the dosage, and the aggregate grading. High dosages of admixture may cause excessive water demand as well as poor workability. Classical tests of workability can not be applied in this case.

9.5.2 Setting stage

(a) Setting
Most synthetic and natural organic polymers soluble in water are retarders.

(b) Air entrainment
Overdosage can cause air entrainment.

(c) Bleeding
Pumping aid admixtures generally reduce bleeding.

9.5.3 Hardening stage

Strength   At early ages a reduction of strength can be expected. This reduction depends on the dosage of the admixture and the retardation effect, or excessive air entrainment.

9.6 USE

The so-called thickeners are dosed in the range 0.02–0.5% solid by weight of cement and added in the mixer. They should be dissolved in water prior to addition to ensure uniform distribution in the concrete.

Dispersions of natural gums, styrene copolymers with carboxyl groups and synthetic polyelectrolytes are dosed in the range 0.01–0.10% solid by weight of cement and added at the mixer.

Acrylic emulsions are dosed in the range 0.10–1.50% solid by weight of cement and added at the mixer.

The presence of water reducing admixtures generally requires lower dosages of the pumping aid admixture.

9.7 APPLICATIONS

These admixtures can be used with any concretes.

9.8 SPECIFICATIONS

There are no test methods for measuring pumpability, and only indirect indications of effectiveness of the admixture can be determined, as bleeding or water retentivity.

There are no standards or specifications to regulate the use of pumping aids. Only a few documents with information on use, effects and properties are available.

1. Standards Association of Australia, Committee BD/33, (1975) *Concrete Admixtures. Information for Use in Concrete and Mortar.* Draft Miscellaneous Publication MP 20, Part 2.
2. Browne, R.D. and Bamforth, P.B. (1977) Tests to establish concrete pumpability. *Journal of the American Concrete Institute, Proceedings*, Vol. 74, pp. 193–203.
3. American Petroleum Institute, *ACI Recommended Practice for Testing Oil Well Cements and Cement Additives*, API-RB 10 B, 1077 Section 8.
4. Kempster, E (1969) *Pumpable Concrete*, Building Research Station, Garston, UK, Current Paper CP 29/69.

9.9 REFERENCES

1. Russell, P. (1983) *Concrete Admixtures*, Viewpoint Publications (E &FN Spon), London, pp. 86-8.
2. Ramachandran, V.S. (1984) *Concrete Admixtures Handbook: Properties, Science and Technology*, Noyes Publications, New Jersey, pp. 528-32.
3. Valore, R.C. (1978) Pumpability aids for concrete, *ASTM Special Technical Publications 169B*.
4. Standards Association of Australia Committee BD/33, (1975) *Concrete Admixtures. Information on Thickening Admixtures for Use in Concrete and Mortar*, Draft Miscellaneous Publication 20, Part 2.

10

Shotcrete admixtures

D. Dimic

10.1 DEFINITION

Shotcrete admixtures, also called quick accelerators or stiffening admixtures, accelerate the stiffening process in fresh concrete.

10.2 INTRODUCTION

Shotcrete admixtures make possible the spraying of concrete on vertical and overhead surfaces by quick stiffening in order to build up thick layers and to reduce rebound. They may also be used for special applications requiring rapid setting and early strength development, like sealing leaks in below-grade structures, for patchwork and emergency repairs.

In the definition of admixtures according to RILEM TC 84-AAC, provision is made that the dosage shall not exceed 5% by weight of cement. Shotcrete admixtures generally exceed this dosage, but are nevertheless included here.

10.3 COMPOSITION

The most common substances used as shotcrete admixtures are sodium and potassium aluminates, hydroxides and carbonates [1-9], triethanolamine [10], ferric sulphate [11], fluorosilicates [4], sodium fluoride and alkali silicates [4]. Commercial shotcrete admixtures are generally mixtures of various proportions of different ingredients.

Often the admixtures are combined with one or more other materials in order to obtain specially desired modifications. They may include wetting agents, stabilising agents for liquids, corrosion inhibitors and others. Recently some new formulations based on inorganic neutral salts and some others based on neutral organic compounds have been developed [3]. A new shotcrete admixture based on calcium aluminate minerals was developed recently [12, 13, 14].

10.4 MECHANISM OF ACTION

Shotcrete admixtures promote the stiffening of the cement matrix mostly through the acceleration of the reaction between C_3A and gypsum [2, 3, 10]. Most of them probably act by precipitating as insoluble hydroxides or other salts. By forming larger quantities of ettringite at early age, some of them also modify the hydration of C_3S [3]. The rapid reaction of C_3A disturbs the harmonious reaction between the calcium silicates and water [15].

10.5 MAIN FACTORS AFFECTING THE MECHANISMS OF ACTION

There are many factors that affect the action of shotcrete admixtures, e.g.

Mix design

*   cement (type, brand, composition, content etc),
*   the presence of other admixtures – air entraining agents, finely divided mineral additions (inert, hydraulic or pozzolanic reactive solids),
*   the water/cement ratio.

Conditions prevailing during the shotcreting

*   admixture dosage,
*   mix temperature,
*   ambient temperature,

- prehydration of the cement,
- proper addition of the admixture.

Differences between cements, types and brands can vary enough to cause changes in the reaction with the shotcrete admixture. Cements with a higher content of C_3A, as well as more finely ground cements, react faster. Usually the shotcrete admixture is formulated for a Portland cement of average mineralogical composition. However, the formulation of the shotcrete admixture can be varied to suit the chosen cement [8]. Very-low-C_3A sulphate-resistant Portland cement requires a specially formulated shotcrete admixture [16, 17].

Before the application of any shotcrete admixture, a compatibility test must be carried out, and the optimum proportions must be determined at the site with the materials, mix and temperature prevailing. One compatibility test for a shotcrete admixture and Portland cement combination has been proposed by ASTM Sub Committee C 09.03.08.07, Admixtures for Shotcrete [18]. This test will determine the compatibility of the admixture–cement combination, but will not necessarily indicate the time of set this combination will produce in actual shotcreting, or give the optimum proportion of the admixture.

10.6 EFFECTS

10.6.1 Fresh stage

Shotcrete admixtures should contribute to good cohesion and produce a dry gritty mix, so that thicker layers of shotcrete can be built up on overhead or vertical surfaces during a single shift.

10.6.2 Setting stage

Shotcrete admixtures provide rapid stiffening or setting. At optimum dosages the initial setting time may be reduced to 3–4 minutes, while the final setting time can vary between 10 and 15 minutes. More rapid 'flashset' setting times can be achieved by careful selection of a suitable combination of cement and admixture. Fig. 1 gives an example of the

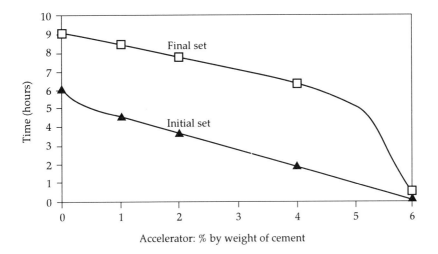

Fig. 10.1 Influence of dosage of a caustic shotcrete admixture on setting time of a shotcrete mix [19] (ASTM C 403 Test Method)

influence of an increasing admixture dosage on the setting time of a plain shotcrete for a particular, highly caustic shotcrete admixture [19].

10.6.3 Hardening stage

Beside the rapid stiffening and setting of the concrete mix, another beneficial effect of the use of a shotcrete admixture may be more rapid early strength development. The maximum rate of strength increase is reported to be within the first few hours [19, 20].

10.6.4 Hardened stage

The development of the physical and mechanical properties of shotcrete, such as strength, shrinkage, permeability and durability, is different from that of ordinary concrete.

(a) Strength
Some decrease in the mechanical strength of the concrete at later ages can be observed when shotcrete admixtures are used. The 28 day strength is usually lower by 20 to 25% at optimum dosages compared

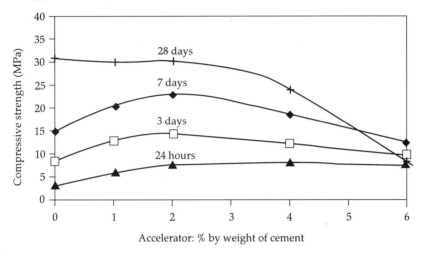

Fig. 10.2 Influence of dosage of a caustic shotcrete admixture on compressive strength development of a plain shotcrete mix [19]

to the reference concrete mix. At higher dosages some shotcrete admixtures cause a strength reduction of up to 50% [19, 20, 21]. Some factors that may cause low strength are: formation of C-S-H with a higher C/S ratio; disturbed C_3S hydration; very rapid setting followed by greater heat development; and a more porous structure [1]. Fig. 2 gives an example of the influence of increasing admixture dosage on the compressive strength development of a plain shotcrete for a particular, highly caustic shotcrete admixture [19]. Some new formulated admixtures are reported to have a less detrimental effect on long-term strength development. In some cases, even no reduction is reported [13].

(b) Shrinkage
It has been found that the drying shrinkage of shotcrete increases when a shotcrete admixture is used.

(c) Durability
There are indications that shotcrete admixtures can adversely influence the durability of shotcrete. Resistance to freezing and thawing is decreased [22]. In the case of alkali–aggregate reactivity, admixtures based on alkali substances can increase the reaction.

(d) Modulus of elasticity

In general, the relationship between compressive strength and modulus of elasticity is similar to that of plain concrete. If adequate curing is provided, values of modulus increase with age [1].

10.7 USE

Shotcrete admixtures are available either in dry or liquid form. Both liquid and powder admixtures can be used with dry-mix and wet-mix process shotcrete equipment [13].

Powder admixtures are generally used at addition rates of 2 to 10% by weight of the cement. Dispensing powder admixture is difficult, so suitable apparatus for dosage of the admixture to the concrete is required. In order to reduce dust-raising and the generation of static electricity caused by the dry mixture passing through the hose, it is desirable to use aggregate with a moisture content of 3 to 5%. However, if the cement remains for a certain time in contact with damp aggregate, partial hydration will take place. This will cause delaying of the setting time of the cement and a reduction in the acceleration effect of the admixture [18].

Liquid admixtures are usually batched by volume ratios, with the admixture : water ratios ranging from 1 : 1 to 1 : 20 [1]. Liquids can be accurately dispensed by positive flow pumps into the gauge water. This provides a good overall distribution.

Most commercially available shotcrete admixtures contain chemicals that are caustic and require caution in handling. They can cause reactions on the exposed skin, causing mild to severe burns. With liquid admixtures, burns can occur immediately. The material must be kept from getting in the eyes and mouths of workers. Personal protection equipment is required for all those in the vicinity of the shotcrete machine and nozzle.

A new method has been developed, to solve technological and environmental problems, which no longer relies on chemical agents, but allows the wet-mix shotcrete to harden immediately after placement as a result of a physical process [23].

10.8 SPECIFICATIONS

There are no standards in existence for shotcrete admixtures. There are different recommendations for shotcrete admixtures [2, 24, 25, 26]. In general the following recommendations are fairly typical for shotcrete with an admixture:

1. The initial and the final set should occur within 3 and 12 minutes respectively.
2. The rate of strength development should be such that values of the order of 4.0 MPa are attained at 8 hours, and 10 Mpa at 24 hours,
3. The decrease in 28 day strength should be less than 30% compared to the concrete without an admixture.

10.9 REFERENCES

1. Ramachandran, V. S., (1984) *Concrete Admixture Handbook: Properties, Science and Technology*, Noyes Publications, USA.
2. Littlejohn, G. S., (1980) Wet process shotcrete, *Proceedings of Symposium on Sprayed Concrete (CI 80)*, London, Construction Press, pp. 18–35.
3. Burge, Th. A., (1982) *Erstarrungnsbeschleuniger fur spitzbeton*, Fachtagung "Spitzbeton" des Schweizerischen Ingenier und Architektur – Vereins (SIA), Zurich.
4. Satalkin, A. V. and Solntsev, V. A., (1967) A new addition to accelerate the seizing process and hardening of spitzbeton, *Proceedings of RILEM-ABEM, International Symposium on Admixtures for Mortar and Concrete*, Brussels, Report III/8, pp 105-13.
5. American Concrete Institute Committee 212, (1985) *Admixtures for Concrete*, ACI 212.3R.-89.
6. Babthev, G. N. and Nikolova, A. B., (1986) Chemical additives for accelerating the setting and hardening of gunite, *Zement –Kalk –Gips*, No. 6, pp. 263-6.
7. Niel, E. M. M. G., (1969) Influence of alkali-carbonate on the hydration of cement, *Proceedings of Fifth International Symposium on the Chemistry of Cement*, Tokyo, V.III, pp. 472–86.
8. Kusterle, W. A., (1983) *Optimierung der Komponenten fur*

Spitzbeton, Universitat Innsbruck Dissertation, Vol. I, II, III.

9. Valenti, G. L. and Sabatelli, V., (1980) The influence of alkali carbonates on the setting and hardening of Portland and pozzolanic cements, *Silicates Industriels*, Vol. 45, pp. 237–42.

10. Ramachandran, V. S., Influence of triethanolamine on the hydration of tricalcium aluminate, *Cement and Concrete Research*, 3, pp. 41–54 (1973).

11. Rosenbery, T. J., (1967) Investigation of trivalent iron salts as admixtures accelerating the hardening of concrete, *Proceedings RILEM-ABEM International Symposium on Admixtures for Mortar and Concrete*, Brussels, Report III-IV/10, pp. 169-80.

12. Nakagawa, K. and Udagawa, H., (1988) *Use of Ettringite*, International Meeting on Advanced Materials, MRSA, Sunshine City, Tokyo, May 30–June 3.

13. Nakagawa, K, (1986) *Accelerating Agents for Shotcrete and Hardening Accelerators for Concrete Repair*, The First Annual Lecture Meeting of New Materials, Methods and Machines, JSCE.

14. Nakagawa, K., Gomi, H. and Udagawa, H., (1990) Shotcrete admixture based on calcium aluminate minerals, *International RILEM Symposium on Admixtures for Concrete*, Barcelona, Spain, paper not included in the proceedings.

15. Ramachandran, V. S., (1994) Interactions of admixtures in the cement–water system, *Application of Admixtures in Concrete*, A. M. Paillere, editor, RILEM TC 84 AAC - Final Report, E & FN Spon, London.

16. Huber, H., (1981) Neue entwicklungen bei spitzbeton, *Zement und Beton*, No. 3, pp. 103–5.

17. Singh, M. N., Seymour, A. and Bortz, H., (1973) The use of special cement in shotcrete, *Use of Shotcrete for Underground Structural Support*, ACI SP-45, pp. 200–31.

18. Shutz, R. J., (1981) Factors influencing dry-process accelerated shotcrete, *Concrete International*, Vol. 3, No. 1, pp. 75–9.

19. Morgan, D. R., (1981) Recent developments in shotcrete technology – a materials engineering perspective, *The World of Concrete '88*, Las Vegas, USA, p. 54.

20. Sereda, P. J., Feldman, R. F. and Ramachandran, V. S., (1980) Influence of admixtures on the structure and strength development, *Proceedings Seventh International Congress on the Chemistry of Cement*, Paris, Part VI, pp. 32–44.

21. McCurrich, L. H., Lammiman, S. A. and Hardman, M. P., (1980) Special purpose admixtures, *Concrete International (CI 80)*, Proceedings International Congress on Admixtures, London, Construction Press, pp. 73–87.
22. Reading, T. J., (1981) Durability of shotcrete, *Concrete International*, Vol. 3, No. 1, pp. 27–33.
23. Wichern, R., (1991) Practical experience and new developments with regard to repair shotcrete and mortars, *Structural Repair, Large Scale Renewal, Preventive Protection and Special Repair Methods of Reinforced Concrete Structures*, Universitat Innsbruck, pp. 119–124.
24. Humphries, E. F., (1980) Specifications and codes of practice for sprayed concrete, *Proceedings of Symposium of Sprayed Concrete (CI 80)*, pp. 8–17.
25. Plotkin, E. S., (1981) Tunnel shotcrete lining, *Concrete International*, Vol. 3, No. 1, pp. 94–7.
26. Japan Society for Civil Engineering, (1986) *Proposed Specifications for Shotcrete Admixtures*, pp. 103–8.

APPENDIX

Guide for use of admixtures
in concrete

Materials and Structures, 1992, **25**, 49–56

RILEM TECHNICAL COMMITTEES
COMMISSIONS TECHNIQUES DE LA RILEM

84-AAC: APPLICATION OF ADMIXTURES FOR CONCRETE

Guide for use of admixtures in concrete

A. M. PAILLÈRE
Laboratoire Central des Ponts et Chaussées, Paris, France

M. BEN BASSAT
Technion, Haifa, Israel

S. AKMAN
Technical University, Istanbul, Turkey

1. DEFINITION

Admixtures of concrete mortar or paste are inorganic (including minerals) or organic materials in solid or liquid state, added to the normal components of the mix, in most cases up to a maximum of 5% by weight of the cement or cementitious materials.

The admixtures interact with the hydrating cementitious system by physical, chemical or physico-chemical action, modifying one or more properties of concrete, mortar or paste in the fresh, setting, hardening or hardened state.

Materials such as fly ash, slag, pozzolanas or silica fume which can be constituents of cement and/or concrete, also products acting as reinforcement, are not classified as admixtures.

2. EFFECTS OF ADMIXTURES FOR CONCRETE IN MOST FREQUENT USE

2.1 Main effect exerted in the fresh state

Water reducing agents either increase the slump of freshly mixed mortar or concrete without increasing the water content or maintain workability with a reduced amount of water.

High-range water reducing agents (super-plasticizers) are admixtures which procure a considerable increase in the workability of mortars and concretes at constant water to cement ratio. The duration of the effect is generally temporary and variable. Mortars and concretes of constant workability can be made with smaller amounts of water, saving more than 12%, without undue retardation, excessive entrainment of air or detrimental bleeding.

0025-5432/92 © RILEM

2.2 Main effect exerted at the setting stage

Set-retarders are products which regard the initial rate of reactions between cement and water. They increase the time taken by mortars and concretes to pass from the plastic to the solid state.

Set-accelerators are products which accelerate the initial rate of reactions between cement and water. They reduce the time taken by mortars and concrete to pass from the plastic to the solid state.

2.3 Main effect exerted in the hardening stage

Hardening accelerators are admixtures which accelerate the development of early strength in the concrete, or mortar or grout, whether or not they affect the setting time.

2.4 Main effect exerted in the hardened stage

Air-entrainers are products which cause the formation of minute, uniformly distributed micro-bubbles of air in the concrete or mortar. They remain after hardening. Air-entraining gives hardened mortars and concretes freeze–thaw resistance.

3. ADMIXTURE INTERACTION IN THE WATER–CEMENT SYSTEM

There are often several different types of interaction between cements and admixtures. They vary according to the nature of the admixture. Chemical reactions which may develop at the beginning of contact between cement and water are complex. Not all the mechanisms of admixture action are completely understood. These mechanisms can involve phenomena of physical or chemical adsorption and chemical reactions with certain cement constituents.

Accelerators (organic or inorganic) generally act by accelerating either the hydration of the aluminates (in combination with gypsum) or the hydration of silicates contained in the cement.

Retarders such as sugar, hydroxycarboxylic acids or lignosulphonates promote reaction by absorption or by formation and fixing of compounds at the surface of aluminates or silicates. It is often accepted that retardation is also due to the formation of a protective layer on the surface of cement grains which slows down initial hydration reactions.

Admixtures which modify the rheological properties of cement (superplasticizers or water-reducing agents) generally act by fixing on the surface of cement grains, modifying the surface state. This produces 'dispersion' or deflocculation of cement grains in the fresh concrete. This type of surface interaction can explain certain secondary effects, as in the case of certain lignosulphonates, which are not only water-reducing agents and plasticizers but also retarders.

The control of *setting* depends particularly on an equilibrium, which is quite difficult to ensure, between the reactivities and the respective contents of the different sulphates and binder. This equilibrium may be modified in certain cases by interaction with the admixture. This has been observed sometimes with certain combinations of cement–lignosulphonate or gluconate cements. For this reason it is appropriate to undertake preliminary tests with the cement and the admixture chosen for a specific structure, preferably by using an extra amount of admixture, to make sure there

is no excessive modification of the cement setting characteristics.

Air-entraining agents promote adsorption reactions at the water–air interface (giving rise to entrapped microscopic air bubbles) and sometimes on certain solid surfaces of concrete containing fly ash with a high calcium content. Air-entrainers are adsorbed on them, thus higher than normal amounts of the admixture would be needed.

4. EFFECTIVENESS OF ADMIXTURES

The effectiveness of each admixture may vary depending on its concentration in the concrete and various constituents of the concrete, particularly the cement.

Each class of admixture is defined by its main function. It may have one or more secondary functions and its use may result in side-effects. The *main function* is characterized by the effect(s) it has on the required properties of the concrete. The *secondary function(s)* concern(s) effects that are in most cases independent of the main function. An example is as follows:

Main function – water reducing
Secondary function(s) – set-retarding, air-entraining

The use of an admixture may result in effects on the properties or behaviour of the concrete that, even though unsought, are inevitable. These effects are known as side-effects (examples include a loss of mechanical strength, set-retarding, increased shrinkage, etc.). These side-effects must be detected or confirmed, often by means of preliminary investigations specific to each case. Table 1 indicates the

Table 1 Main and secondary effects possible for every type of admixture in frequent use. These effects are not a general rule but depend on the admixture formulation. Some formulations containing chloride may be deleterious to reinforcement

Effect[a]	Type of admixture					
	Water-reducers and high-range water-reducers	Super-plasticizers	Setting accelerators	Setting retarders	Hardening accelerators	Air-entrainers
Water reduction (see Tables 2 and 3)	× ×	×				×
Workability increased (see Tables 2 and 3)	×	× ×				×
Setting (see Tables 2 and 4)	×	×	× ×	× ×	×	
Hardening (see Tables 2, 4 and 5)	×		×	×	× ×	
Air entrainment (see Tables 2 and 6)	×					× ×
Strength (see Tables 2 and 5)	× ×	×			× × (early age)	×
Durability (see Tables 2 and 6)	×					× ×

[a] × × = main effect, × = secondary effect.

main functions and possible secondary effects of non-chloride admixtures.

5. THE USE OF ADMIXTURES

Depending of the materials available (cements, aggregate), limitations in mixing or placing, ambient conditions and placement problems associated with the project requirements, it is impossible to produce satisfactory concrete without the use of an admixture. Consequently, admixtures can be considered as constituents of the concrete, just as are cement, aggregates and water. They are often used to optimize the cost-effectiveness of a concrete mix.

It should be emphasized that admixtures provide additional means of controlling the quality of the concrete by modifying some of its properties, but they cannot, whatever the conditions, correct poor quality of materials, unsatisfactory proportioning of the concrete and inappropriate placement procedures.

6. CONCRETE PROPERTIES AFFECTED BY ADMIXTURES

6.1 The fresh stage

This is the period of time during which concrete already made can be handled, transported, poured and finished before initial setting takes place.

Unit mass is the mass of fresh concrete per unit

Table 2 Effect of types of admixture in frequent use on concrete properties which can be modified

	Type of admixture					
	Water-reducers	High-range water-reducers	Super-plasticizers	Accelerators	Retarders	Air-entrainers
Dosage by weight of cement (%)	< 0.5	0.5 to 3.0	0.5 to 3.0	< 0.5	< 0.5	0.02 to 0.10
Method of introduction into mix	Into mixing water	Into mixing water	Before placing	Into mixing water	Into mixing water	Into mixing water
Air content	(0) or (+)	(0) or (+)	(0) or (+)	(0)	(0) or (+)	(+)
Fresh stage						
Unit mass (kg m⁻³)	(+)	(+)	(+)	(0)	(0) or (+)	(−)
Workability	Can modify setting times					
Water demand	(−)** (0)*	(−)**	(0)*	(0)	(0) or (−)	(−)** (0)*
Consistency	(0)** (+)*	(0)**	(+)*	(0) or (−)	(0) or (+)	(0)** (+)*
Consistency loss	(0)** (+)*	(+)**	(+)*	(0) or (+)	(0)	(0)
Pumpability	(+)** (+)*	(+)**	(+)*	(0)	Can modify water demand	Improvement
Segregation	(−)** (0)*	(−)**	(+)* or (0)	(0)	(0) or (+)	(0) or (−)
Setting and hardening stage						
Setting time						
Initial	(0) or (+)	(0) or (+)	(0) or (+)	(−)	(+)	(0) or (+)
Final	(0) or (+)	(0) or (+)			(+) or (0)	
Strength development	(+)** (0)*	(+)**	(0)*	(+)	(−)	(0)
Bleeding	(−)** (+)*	(−)**	(0) or (−)	(−)	(0) or (+)	(−)
Plastic shrinkage	(0) or (−)	(+)	(+)	(−)	(+)	(0) or (−)
Hardened stage						
Strength						
Flexural < 3 days	(0) or (+)** (0)* or (−)*	(+)**	(−)* or (0)*	(+)	(0) or (−)	(−)*
> 28 days	(+)** (0)*	(+)**	(−)* or (0)*	(0) or (−)	(0)	(−)*
91 days	(+)** (0)*	(+)**		(0) or (−)	(0) or (+)	(−)
Compressive < 3 days	(0) or (+)** (0)* or (−)*	(+)**	(−)* or (0)*	(+)	(0) or (−)	(−)*
> 28 days	(+)** (0)*	(+)**	(−)* or (0)*	(0) or (−)	(0)	(−)*
91 days	(+)** (0)*	(+)**		(0) or (−)	(0) or (+)	(−)*
Modulus of elasticity	(0) or (+)	(0) or (+)	(0) or (+)	(0)	(0)	(−)
Durability						
Capillary absorption/ permeability	(−)** (0)*	(−)**	(0)*	(0)	(0) or (−)	(+)
Freeze–thaw resistance	(+)** (0)*	(+)** (0)**	(0)*	(0)	(0) or (+)	(+)
Thermal expansion	(0)	(0)	(0)	(0)	(0)	(−)
Creep	(0) or (+)	(0) or (−)**	(0)*	(0)	(+)	(+)
Shrinkage	(0) or (+)	(0) or (+)	(0)*	(0)	(0) or (+)	(0) or (+)
Corrosion of steel	(−)** (0)*	(−)**	(0)*	(0)	(0) or (+)	(−)** (−)*

(0) No effect, (+) increase, (−) decrease, *constant W/C as reference, **constant slump as reference.

volume. This volume is absolute or apparent. The absolute volume is free of air. The apparent volume includes the air content.

Workability is the combination of properties of freshly mixed concrete which enables it to be placed, compacted and finished easily without loss of homogeneity. It includes the following:

1. *Consistency:* the relative ability of freshly mixed concrete to flow.

2. *Plasticity:* the property of freshly mixed concrete which determines its resistance to deformation or ease of moulding.

3. *Cohesion:* the property of a freshly mixed concrete to preserve its homogeneity.

4. *Consistency loss:* the reduction in consistency that occurs with time in relation to an initial measurement taken from the same batch.

5. *Pumpability:* the ability to convey the fresh concrete by pressure through either a rigid pipe or a flexible conduit.

6. *Compactability:* the ability of freshly placed mortar or concrete to reduce its initial volume to the minimum practical space by vibration, centrifugation, tamping or some combination of these, to mould within forms or moulds and around embedded parts and reinforcement, and to eliminate voids other than entrained air.

7. *Finishability:* the ability of freshly placed concrete to be levelled, smoothed or to have its surface worked to provide the desired appearance and serviceability.

Segregation is the non-uniform distribution of solid constituents of a fresh concrete.

Frost resistance. Deterioration occurs if the freezing takes place during setting and/or in the green concrete.

Table 3 Non-chloride frequently used admixtures modifying the workability of concretes and mortars

	Type of admixture			
	Water-reducers	High-range water-reducers	Superplasticizers	Air-entrainers
Main chemical family	Lignosulphonates, hydroxycarboxylic acids, carbohydrates	Melamine formaldehyde condensate, sulphonated naphthalene formaldehyde condensate, modified lignosulphonates, acrylic (graft) copolymer, amino aromatic sulphoneic acid, phenol formaldehyde condensate		See Table 6
Main effects				
Effect on workability	At constant workability water reduction >6.5%[a]	At constant workability water reduction >12%[b]	At constant water to cement ratio, highly fluidizing effect, slump >8 cm	Slightly water reducing at constant workability
Effect on strength (1, 3, 7, 28 and 91 days)	Higher than the reference: increase due to water reduction (10% minimum)	Higher than the reference: increase due to water reduction (15% minimum)	Compared to reference at the same water to cement ratio, small strength decrease is possible	Less than the reference depending on the amount of air entrained (maximum 20%)
Secondary effects				
Favourable	(i) Increase of compactness, decrease of permeability (ii) Possible improvement of resistance to chemical aggressive environments for the same cement contents	Very high strength concretes can be achieved by adopting very low water to cement ratios		(i) Improvement of surface aspect (ii) Improvement of freezing resistance
Unfavourable	(i) Setting retardation with certain formulations (ii) Possible increase of shrinkage (iii) Stiffening by possible admixture–cement interaction	(i) Slump loss except for special formulas and depending on the temperature (ii) Possible shrinkage increase	(i) Segregation when excess of water or admixture or bad grading (ii) Efficiency very sensitive to fines content of cement and sand (iii) Slump loss except for special formulas and depending on temperature (iv) Possible shrinkage increase	Efficiency varying as a function of fine elements, mixing and method of placing concrete

[a] 10% in Japan.
[b] 16% in Japan.

Table 4 Non-chloride frequently used admixtures modifying the setting time and hardening of cement

	Type of admixture		
	Setting accelerators	Setting retarders	Hardening accelerators
Main chemical family	Calcium nitrites, nitrates, thiosulphates–formates, triethanolamine, etc.	Gluconate–salicylic acid–Ca lignosulphonates, sodium boroheptonate, etc.	Calcium nitrites, nitrates, thiosulphates, formates, triethanolamine, etc.
Main effects			
Effect on setting time	Acceleration very variable as a function of admixture content, type of cement and temperature	Retardation very dependent on admixture content, type of cement and temperature	Acceleration very variable as a function of admixture content, type of cement and temperature
Effect on strength			
Before 3 days	Increased at 1 or 2 days	Decreased at 1 or 2 days	Increased (until 7 days)
After 28 days	Slightly decreased (the greater the reduction in strength, the greater the setting acceleration)	Slightly increased	Unmodified or slightly decreased
Secondary effects			
Favourable	In cold weather increase of heat of hydration during first few hours	Water reduction	In cold weather increase of heat of hydration during first few hours
Unfavourable	(i) In hot weather (ii) Increase of heat of hydration during first few days (iii) Slight increase of shrinkage (iv) Possible stiffening (v) Alkali–aggregate reactions possible if admixture contains alkalis	Shrinkage unmodified or slightly increased	(i) In hot weather (ii) Increase of heat of hydration during first few days (iii) Slight increase of shrinkage (iv) Possible stiffening (v) Alkali–aggregate reactions possible if admixture contains alkalis

Table 5 Non-chloride frequently used admixtures modifying the strength

	Type of admixture				
	Water-reducers	High-rate water-reducers	Hardening accelerators	Setting retarders	Air-entrainers
Main chemical family	See Table 6	See Table 3	See Table 4	See Table 4	See Table 6
Main effects					
Effect on strength					See Table 6
Before 3 days	Higher than reference at all ages, increase due to water reduction for same workability	At constant workability much higher than the reference, at all ages	Hardening possible between 2 and 10°C	Lower than reference	
After 28 days		Utilization in high-strength concretes	Possible reduction of final strength, depending on nature of admixture	Possible increase of final strength	
Secondary effects					
Favourable	(i) Increase of compactness (ii) Improvement of resistance to chemical aggressive agents (iii) Improvement of surface aspect	(i) Increase of compactness (ii) Improvement of resistance to chemical aggressive agents (iii) Improvement of surface aspect	Setting acceleration for certain formulas		See Table 6
Unfavourable	(i) Setting retardation with certain formulations (ii) Possible increase of shrinkage (iii) Stiffening by possible cement–admixture interaction	(i) Possible increase of shrinkage (ii) Possible incompatibility with certain air-entrainers	Possible increase of shrinkage	Possible increase of shrinkage	

6.2 The setting stage

This is the period of time during which fresh concrete in place starts setting, up to before hardening initiates.

Setting is a condition defined as occurring at the instant when concrete becomes unsuitable for placement.

Heat of hydration is heat evolved by chemical reactions associated with cement hydration during its setting and hardening.

Plastic shrinkage is the reduction in volume during setting and/or hardening of concrete due to excessive and rapid evaporation of bleeding water, setting up hydrostatic tension in capillary pores within the mass.

Bleeding is the displacement of water from freshly placed concrete caused by settlement or flocculation of solid materials within the mass. It is also called 'water gain'.

6.3 The hardening stage

This is a physico-chemical phenomenon causing the particles of hydraulic binders to combine and form the solid skeleton as a result of progressive hydration of the anhydrous components, resulting in a progressive strength gain after setting. *Heat of hydration* is evolved and *strength development* takes place.

Shrinkage is the reduction in volume occurring at the hardening and hardened stage of the concrete, caused by physical changes and/or chemical reactions. It is a function of time but not of external causes like temperature and load.

6.4 The hardened stage

This is the period of service life of concrete after a specified attained strength. It is characterized by *unit mass* and by *strength*, measured as follows:

1. *Compressive strength:* the maximum resistance of a concrete to axial loading.

2. *Flexural strength:* the calculated ultimate stress in the outer fibre of a concrete beam specimen subjected to bending.

2a. *Direct tensile strength:* the maximum unit stress which a concrete is capable of resisting under axial tensile loading.

3. *Bond:* the adhesion and grip of concrete to reinforcement.

Modulus of elasticity is the ratio of the normal stress, below the limit of proportionality of concrete, to the measured instantaneous unit deformation.

Durability is the ability of concrete to resist weathering action, chemical attack, abrasion and other conditions of service. It comprises the following factors:

1. *Porosity:* the ratio, usually expressed as a percentage, of the volume of voids in a material to its total

Table 6 Non-chloride frequently used admixtures modifying the resistance against aggressive media (durability)

	Type of admixture	
	Water-reducers and/or high-range water-reducers	Air-entrainers
Main chemical family	See Table 3	Salts of wood resins, fatty and resinous acids and their salts, organic salts of sulphonated hydrocarbons, etc.
Main effects		
Freeze–thaw resistance	Improvement due to increased compactness and mechanical strength by using less water but maintaining workability	Highly increased
Resistance to aggressive atmospheric agents (CO_2, sea weathering)	Improvement due to increased compactness and mechanical strength by using less water but maintaining workability	?
Resistance to chemical aggressive agents (sulphates, sea water, etc.)	Improvement due to increased compactness and mechanical strength by using less water but maintaining workability	Possible improvement due to lowering of capillary absorption
Water penetration	Decrease due to increase of compactness by water reduction	Possible decrease due to lowering of capillary absorption
Secondary effects		
Favourable	Increase of mechanical strength due to water reduction for same workability	Improvement of surface aspect
Unfavourable		Possible decrease of mechanical strength (maximum 20%)

volume, including the voids. It affects

(a) *capillary absorption:* the suction of water into the capillary pores of solidified concrete when one face has been placed in contact with water for a given period, and

(b) *permeability:* the capacity of a hardened concrete to allow flow of a fluid through it under a hydraulic gradient.

2. *Freeze–thaw resistance:* resistance of the hardened concrete to freezing and thawing cycles.

3. *Attack by aggressive solutions:* physico-chemical reaction between cement paste and soluble sulphates in surrounding media, primarily with tricalcium aluminate hydrates.

4. *Carbonation:* reaction between carbonic acid and the lime in the hydrated cement paste to produce calcium carbonate.

5. *Alkali–aggregate reaction:* a reaction between the alkali oxides (sodium and potassium) in some Portland cements and a reactive silica component in certain aggregates, causing abnormal expansion and cracking of concrete in service.

6. *Efflorescence:* a deposit of salts, usually white, crystallized on the surfaces of building materials and caused by capillary rise or the leaching associated with water penetration.

Corrosion of reinforcement is oxidation and deterioration of the reinforcement in concrete by an electro-chemical process. It involves *ionic diffusion*, the migration of ions through materials due to environmental changes.

Volume change is either an increase or decrease in volume:

1. *Thermal expansion* is expansion caused by increase in temperature.

2. *Creep* is time-dependent deformation due to sustained load, obtained by subtracting from the total deformation the instantaneous deformation at the instant of loading and the drying shrinkage during this period.

3. *Shrinkage.*

Miscellaneous effects include biological attack due to micro-organisms.

The properties that may be modified by the main types of admixtures are summarized in Table 2. Tables 3–7

Table 7 Information sources

Characteristics to be known	Verification or information provided by			
	Verification procedure of conformity with standards	Laboratory testing	Suitability tests	Others
Acceptance		×	×	
Quality control	×			
Efficiency main function	×		×	Technical assistance of manufacturer and/or supplier
Secondary functions	×	× Adapted to importance of work	× Adapted to importance of work	
Secondary effects within the limits of the standard requirements	×	× Depending on specific character of work		
Secondary effects possibility affecting durability	× Standard requirements	× Depending on specific character of work		
Critical characteristics of product	×		×	
Admixture proportioning appropriate to the job site		× Depending on type of work	×	
Interaction with cement used at job site		×	×	Technical assistance of laboratory of cement manufacturer

indicate the contributions of the main types of admixture according to the modification required.

7. PRECAUTIONS IN THE USE OF ADMIXTURES

It is essential when using admixtures to verify the following:

1. The expected effect by means of *testing*. The effect obtained should not be too sensitive to small variations in the amount of the admixture used, or in the amounts of the other concrete constituents.

2. The extent of *side-effects*. The important points that should be checked are summarized in Table 7.

3. That the effectiveness of admixtures is not affected by the *heat curing regime* to which the concrete may be exposed. In this case it is advisable to follow the manufacturer's recommendations and to verify the effect by laboratory and field tests.

ACKNOWLEDGEMENTS

This guide was drafted within the framework of RILEM Technical Committee 84-AAC: Application of Admixtures for Concrete, comprised as follows:

Chairman: Dr R. Rivera Villareal, Universidad Autonoma de Nuevo Leon, Mexico.

Secretary: Dr A. M. Paillère, Laboratoire Central des Ponts et Chaussées, France.

Members: Dr S. Akman, Technical University, Istanbul, Turkey; Dr F. Alou, EPFL, Lausanne, Switzerland; Dr M. Ben Bassat, Technion, Haifa, Israel; Dr S. Biagini, MAC, Italy; Consult. Eng. E. Decker, Roanoke, USA; Dr D. Dimic, Institute for Research and Testing of Materials and Structures, Ljubljana, Yugoslavia; Dr F. Massazza, Italcementi, Italy; Dr Shigeyoshi Nagataki, Tokyo Institute of Technology, Japan; Dr V. S. Ramachandran, National Research Council, Canada; Dr E. Vazquez, Universidad Politecnica de Catalunya, Spain.

Index

Materials and Structures

RILEM's journal, *Materials and Structures*, is published by E & F N Spon on behalf of RILEM. The journal was founded in 1968, and is a leading journal of record for current research in the properties and performance of building materials and structures, standardization of test methods, and the application of research results to the structural use of materials in building and civil engineering applications.

The papers are selected by an international Editorial Committee to conform with the highest research standards. As well as submitted papers from research and industry, the Journal publishes Reports and Recommendations prepared buy RILEM Technical Committees, together with news of other RILEM activities.

Materials and Structures is published 10 times a year (ISSN 0025-5432). Sample copy requests and subscription enquiries should be sent to: E & F N Spon, 2-6 Boundary Row, London SE1 8HN, Tel: (0)71-865 0066, Fax: (0)71-522 9623; or Journals Promotion Department, Chapman & Hall, 29 West 35th Street, New York, NY 10001-2291, USA, Tel: (212) 244 3336, Fax: (212) 563 2269.

RILEM Reports

1 **Soiling and Cleaning of Building Facades**
   Report of Technical Committee 62-SCF. *Edited by L.G.W. Verhoef*
2 **Corrosion of Steel in Concrete**
   Report of Technical Committee 60-CSC. *Edited by P. Schiessl*
3 **Fracture Mechanics of Concrete Structures - From Theory to Applications**. Report of Technical Committee 90-FMA. *Edited by L. Elfgren*
4 **Geomembranes - Identification and Performance Testing**
   Report of Technical Committee 103-MGH. *Edited by A. Rollin and J.M. Rigo*
5 **Fracture Mechanics Test Methods for Concrete**
   Report of Technical Committee 89-FMT. *Edited by S.P. Shah and A. Carpinteri*
6 **Recycling of Demolished Concrete and Masonry**
   Report of Technical Committee 37-DRC. *Edited by T.C. Hansen*
7 **Fly Ash in Concrete - Properties and Performance**
   Report of Technical Committee 67-FAB. *Edited by K. Wesche*
8 **Creep in Timber Structures**
   Report of TC 112-TSC. *Edited by P. Morlier.*
9 **Disaster Planning, Structural Assessment, Demolition and Recycling**
   Report of TC 121-DRG. *Edited by C. De Pauw and E.K. Lauritzen*
10 **Applications of Admixtures in Concrete**
   Report of TC 84-AAC. *Edited by A.M. Paillere*
11 **Interfaces in Cementitious Composites**
   Report of TC 108-ICC. *Edited by J.-C. Maso*

RILEM Recommendations and Recommended Practice

RILEM Technical Recommendations for the Testing and Use of Construction Materials

Autoclaved Aerated Concrete - Properties, Testing and Design
Technical Committees 78-MCA and 51-ALC

RILEM, The International Union of Testing and Research Laboratories for Materials and Structures, is an international, non-profit making, non-governmental technical association whose vocation is to contribute to progress in the construction sciences, techniques and industries, essentially by means of the communication it fosters between research and practice. RILEM activity therefore aims at developing the knowledge of properties of materials and performance of structures, at defining the means for their assessment in laboratory and service conditions and at unifying measurement and testing methods used with this objective.

RILEM was founded in 1947, and has a membership of over 900 in some 70 countries. It forms an institutional framework for cooperation by experts to:

* optimise and harmonise test methods for measuring properties and performance of building and civil engineering materials and structures under laboratory and service environments;
* prepare technical recommendations for testing methods;
* prepare state-of-the-art reports to identify further research needs.
* collaborate with national and international associations in realizing their objectives

RILEM members include the leading building research and testing laboratories from around the world, industrial research, manufacturing and contracting interests as well as a significant number of individual members, from industry and universities. RILEM's focus is on construction materials and their use in buildings and civil engineering structures, covering all phases of the building process from manufacture to use and recycling of materials.

RILEM meets these objectives though the work of its technical committees. Symposia, workshops and seminars are organised to facilitate the exchange of information and dissemination of knowledge. RILEM's primary output are the technical recommendations. RILEM also publishes the journal *Materials and Structures* which provides a further avenue for reporting the work of its committees. Many other publications, in the form of reports, monographs, symposia and workshop proceedings, are produced.

Details of RILEM membership may be obtained from RILEM, École Normale Supérieure, 61, avenue du Pdt Wilson, 94235 Cachan Cedex, France. Tel: Intl + 331 4740 2397. Fax: Intl + 331 4740 0113.

Details of the journal and the publications available from E & F N Spon/Chapman & Hall are given below. Full details of the Reports and Proceedings can be obtained from E & F N Spon, 2-6 Boundary Row, London SE1 8HN, Tel: (0)71-865 0066, Fax: (0)71-522 9623.